BUILDING TH
OF TH

1850–1900

For My Mother, Margaret E. Kerr,
From Whom I Learned Much More Than
Multiplication Tables

Contents

	List of Maps and Illustrations	viii
	List of Tables and Graphs	ix
	Preface	xi
	Orthography and Notes on Abbreviations	xvi
1.	Introduction	1
2.	Getting Started and Subsequent Progress	16
3.	Contractors, Engineers and Petty Contractors: The Varieties and Problems of Management	4
4.	Obtaining Labour	
5.	Work and Working Conditions	
6.	Worker Resistance	
7.	Conclusion	
	Appendix I: How Many?	6
	Appendix II: Tables	11
	Bibliography	227
	Index	249

List of Maps and Illustrations

Map 1 South Asia Selected Physical Features 27

Map 2 Rivers, Cities, and Areas 28

Map 3 Growth of the Railways, 1861–1901 39

Plate 1. Left training bund, Curzon bridge site over the Ganges river at Allahabad, *c.* 1903

„ 2. Pathan workmen, *c.* 1880

„ 3. Bell's patent dredgers, Sher Shah bridge over the Chenab river, *c.* 1889

„ 4. Loading a well to force it to sink, Curzon bridge, *c.* 1903

„ 5. Girder erecting, Landsdowne bridge over the Indus river at Sukkur, *c.* 1888

„ 6. Lime-kilns, Bezwada bridge, *c.* 1891

„ 7. Mortar mills, Sher Shah bridge, *c.* 1889

„ 8. Platelaying in the Peshawar City station, mid-1880s

„ 9. View of the Bhore Ghat and the reversing station, early 1860s

„ 10. Masonry bridge, Bhore Ghat, *c.* 1860

Photographs 1, 2, 9 and 10 come from collections held in the Department of Prints, Drawings and Photographs, Oriental and India Office Collections, The British Library, London, England. The other six photographs come from collections held in the Library of The Institution of Civil Engineers, London, England. The photographs are reproduced with the permission of the The British Library—Oriental and India Office Collections, and the Library of the Institution of Civil Engineers. The author is grateful for the permission.

List of Tables and Graphs

Table 1 Miles open (1853–1900) and under construction (1859–1900) 211–12

Table 2 Estimated yearly construction employment, 1859–1900 213–14

Table 3 European construction employment, 1850–9 215

Table 4 Work people employed daily on the Madras Railway, District 3, for the half-year ending 31 December 1855 216

Table 5 Mean daily employment on the Madras Railway by district, and the total rupee expenditure and the amount of earthwork and masonry completed in each district, 1857 217

Table 6 Mean daily employment in December 1858 on construction divisions of the East Indian Railway, Bengal 218

Table 7 Mean number of workers employed daily from 31 May 1859 to 31 May 1860 on construction divisions of the East Indian Railway, Bengal 219–20

Table 8 Mean daily employment, East Indian Railway in the North-Western Provinces, June 1855 221

Table 9 Mean daily employment, East Indian Railway in the North-Western Provinces, January, April and June 1860 221

Table 10 Mean daily employment on the Great Indian Peninsula Railway construction contracts, 1858–64 222

Table 11 Construction employment, Indian Midland Railway, April–June 1886 223

Table 12 Miscellaneous construction employment figures 224

Table 13 Statements of mean employment per construction mile 225

Table 14 Mean daily employment on the Bhore and Thal Ghat inclines, 1856–65 226

Table 15 Estimated employment at selected great bridges 226

Table 16 Estimated yearly construction employment, 1859–1900 (tables 2 and 16 are identical)

Graph Miles open and under construction, 1869–1900 39

Preface

All too many years ago I began a research project into the history of railway workers in colonial India. My goal was to write a book focused primarily on the employees of the operating lines, in which a discussion of the construction of the railways would be a stage-setting early chapter. The operating-line employees were, I thought and still do, the paradigmatic representatives in colonial India of an emerging industrial proletariat. Influenced by Marx, I was inclined, at that time, to see the unfolding of the labour history of India in terms of the emergence of the industrial proletariat; it was to be the leading class fraction within the development of industrial capitalism as the dominant mode of production in India. Many political consequences, past and future, were to be expected as this teleology materialized. The historical role of the employees of the operating railways was clear. All I had to do was to describe what happened and show how the employees fulfilled their destiny.

So much for plans and teleologies! Years later, perhaps a little wiser but also a little sadder, I have written a very different book. Here I offer an extensive description and analysis of the construction of railways in India from their inception in 1850 to approximately 1900. I seek to describe a complex assembly of labour processes, from the perspective of those who conceived, co-ordinated and controlled the building of India's nineteenth-century railways—in short, I seek to explain how the British got the railways built. My focus has shifted from railway operation to railway construction, and from the workers to those who exercised authority over them.

The shift, however, does not mean that I have abandoned the workers as a research interest or my fundamental allegiance to the cause of labour in its long, complicated and on-going struggle with capital. I have simply come to realize better the significance of management in the labour process. This understanding relates not only to India's railway history but also to my own experience as an active member of a union. Praxis even in the form of white-collar unionism in the privileged setting of a late-twentieth-century Canadian university, can teach a historian something about labour processes. It can offer insights into the workings of an earlier much different labour process, even a process as distant and different as railway construction in nineteenth-century India. One learns, for example, that initiative in the exercise of authority within a labour process almost always rests with the factotums of capital. I now know this experientially, and understand this historically.

A discussion of a labour process without a discussion of those who labour is ridiculous. The British got the railways of India built through the labour power of tens of millions of Indians. Indian construction workers are not presented in this book, as my original plan anticipated, as a brief prologue to the putatively more important operating-line employees. They appear as an important element in their own right in the on-going history of labour in India within the continuing transition to capitalism. Construction workers of all sorts, in India and elsewhere, are a significant part of the working class under capitalism and their relationship with capital may well be more typical and enduring than the industrial labour and machino-facture in which Marx saw his vision of the future.

Thus, as this project developed, my thinking diverged in two major ways from my initial expectations. Firstly, at the level of substance I realized that the story of how and by whom the nineteenth-century railways of India had been constructed was an important aspect of India's colonial and labour history, and this had not been adequately told. I discovered it to be the compelling story of a magnificent, collective, often dangerous, accomplishment by millions of South Asian men, women and children directed and controlled, in marked contrast, by a small number of British engineers, contractors, skilled workmen and colonial bureaucrats. It became a story I wanted to tell both for its intrinsic interest and for its significance, in a number of dimensions, to the history of colonial India.

Secondly, the substance of the history of railway construction work in nineteenth-century India forced me to re-examine some of my initial beliefs about the nature, consequences and direction of labour within the emergence of capitalism. In particular, I came to understand that many forms of the labour process are compatible with capitalism and that there is no necessary connection between the advance of capitalism and what Marx called the transition from formal to real subsumption of labour under capital; from the initial subordination of labour to capital and the extraction of absolute surplus value within a labour process technologically little different from the pre-capitalist era, to the so-called real capitalist labour process characterized by the extraction of relative surplus value in a context of large-scale industrial production based upon technology and science, the extensive division of labour and tight managerial control. Indeed, as I also came to understand better that one could not write about a labour process without a discussion of how that process was supervised and controlled (i.e. management), I realized how effectively compatible a variety of work forms and managerial forms were with the central relationship of capitalism: the profit (surplus value) realized by capital from its purchase of labour power and its use of labour.

These two emergent understandings shape the contents of this book and also help to define the audiences to whom it is primarily addressed. The content is largely a plain-language discussion of the building of India's nineteenth-century railways with a particular focus on the management (broadly defined) of the construction process, the work of construction and the people who performed that work. The approach is largely analytical but the untold story does get told. Most of what I convey can be read without reference to my informing perspective although that perspective does shape my presentation and judgements. Those interested generally in railway history, as well as those interested in an important, hitherto largely unexplored, aspect of the nineteenth-century history of India, will, I hope, find material of interest in this book, including the all-too-few illustrations I was able to include from what is a rich photographic record.

Those who come to this book more as historians of labour and wish to know what it may contribute to the on-going debates about the nature and consequences of the labour process(es) within capitalism, should know that I write in the compelling light cast by volume one of Marx's *Capital*, Harry Braverman's *Labour and Monopoly Capital*, and Michael Burawoy's *The Politics of Production*. These works offer a powerful illumination of the social landscapes in which men and women labour, which helped me to find my way through the story of railway construction in nineteenth-century India.

Indeed, this book remains a stage-setting exercise, for I do not claim to have exhausted, or even touched, many important facets of railway construction in India. In particular, I have only set the framework within which detailed investigations of the construction workers and their experience with proletarianization can be pursued. I also chose early to omit a discussion of railway construction as a famine relief measure, since it would have inflated the size of the railway construction workforce (and that of this book) and introduced an element tangential to the main focus of this study. I believe railway construction under famine relief conditions more properly belongs under the history of the colonial state's efforts to provide famine relief. However, I do not deny that the famine gangs were, in their own way, enmeshed in the construction labour process and that they contributed to building the railways of the Raj.

I have acquired substantial debts during the long period required to research and write this book. Research leaves from the University of Manitoba, a generous research grant from the Social Sciences and Humanities Research Council (SSHRC), a Shastri Indo–Canadian Institute fellowship, and the University of Manitoba–SSHRC Fund small research grants made it possible for me to make the necessary research trips to Britain and India. The staffs of the British Library (Oriental and India Office Collections, the British Museum and the Colindale Newspaper collection), the Library of the Institution of Civil Engineers (London), the National Archives of India (New Delhi), the Maharashtrian State Archives (Bombay) and the Ames Library of South Asia at the University of Minnesota (Minneapolis) proved unfailingly helpful. I am grateful to Mrs. P. Kattenhorn, Department of Prints, Drawings and Photographs, Oriental and India Office Collections, the British Library and to Mr M. Chrimes, the Librarian, the Institution of Civil Engineers, for their help with the photographic record and for their kind permission to reproduce the photographs included in this book.

Along the way numerous individuals provided me with advice, assistance and support. I hesitate to list them for fear that I may inadvertently leave someone out. Nonetheless, to A.K. Arora, Richard Bingle, Michael Chrimes, Frank Conlon, Frank Dadswell, Rajiv Dar, Stephen and Jharna Gourlay, P.S. Gupta, Marg Halmarson for the clear maps that appear in this work, A. Joshi, Fritz Lehmann, P.N. Malik, K. Moorthy and Ram Tiwari, I offer my thanks. Joseph Schwartzberg deserves my special thanks: for the provision of base maps and, more importantly, for thirty-five years of sound advice, knowledge and an open office door.

Others provided me with a critical read of drafts of my typescript. Their suggestions and criticisms saved me from many large and small mistakes and improved the typescript as it slowly emerged through various drafts. In so far that I have failed to heed their advice, I will have to pay the price for my stubbornness. I cannot absolve myself of my responsibility for the errors and imperfections that remain. I thank Sabyasachi Bhattacharya, John McGuire, Hew McLeod, and my colleagues, Henry Heller, John Kendle and Ed Moulton for commenting critically, yet usefully, on my work in progress. The penultimate draft received a careful and helpful review from an anonymous OUP assessor, which resulted in a tighter and better focused final draft. I particularly thank Morris David Morris and Burton Stein whose detailed criticisms of the first extensive draft were invaluable. I also thank Burt and Morris for what now approaches an academic lifetime of mentorship and sponsorship. My debt to them is not easily described. Their help, judicious criticisms and published scholarship have assisted, challenged and enriched me. Finally, I thank Kaye whose support and companionship now stretches over thirty years. Even for another academic it is, I suspect, not easy to be married to someone who always seems to be in his study.

Place Name Orthography, and Notes on Abbreviations

I. Place Name Orthography

Place names in this book are spelt as they appear in *The Imperial Gazetteer of India*, vol. 26: *Atlas* (new revised edn, Oxford: Clarendon Press, 1931). Most of the place names which do not appear in the maps in this book can be located in this volume of the *Imperial Gazetteer*. To facilitate the use of this volume an older orthography has been used.

II. Abbreviations

A number of specific items and categories of items appear often enough in the source notes and the text to justify the use of abbreviations. They are:

1. Morris & Dudley = Morris David Morris and Clyde B. Dudley, 'Selected Railway Statistics for the Indian Subcontinent (India, Pakistan and Bangladesh), 1853—1946–47', *Artha Vijnana*. Journal of the Gokhale Institute of Politics and Economics, 17:3 (September 1975), pp. 187–298.

2. *CEHI* = *The Cambridge Economic History of India*, vol. 2: c. *1757–* c. *1970*, edited by Dharma Kumar (Cambridge: Cambridge University Press, 1983).

3. Technical Paper = This refers to a series begun in 1899 whose full citation would be, for example, India, Director of Railway Construction, Technical Paper no. 71: 'The Bridge of the North Western State Railway over the Chenab at Sher Shah, 17 spans of 206 feet, and the Bridge of the East Coast State Railway over the Kistna at Bezwada, 12 spans of 300 feet', by Francis J.E. Spring (1900) which I cite from the very beginning of the book as Spring, Technical Paper no. 71. Other items in the series are cited as Technical Paper no. x and the rest of the citation in full at the first entry and subsequently by the abbreviation as in the example of the paper by Spring.

4. *MPICE* = *Minutes of Proceedings of the Institution of Civil Engineers.* A list of pieces from this journal appears in the bibliography. A shortened form is used in the footnotes, e.g. Berkley, *MPICE*, 19 (1859–60), p. 608. Obituaries from *MPICE* use a similar short form.

5. *Railway Report* = Beginning with that for 1859, an annual report was presented to Parliament on railways in India with the generic title *Report to the Secretary of State for India in Council on Railways in India.* These annual reports were published as sessional papers. The bibliography provides the command numbers and sessional dates.

6. General

   CE = chief engineer or engineer-in-chief

   Estabs. = Establishments

   Progs. = Proceedings

   PWD = Public Works Department

   RR = In a footnote should be read railway or railways

7. Railway Companies

   BB&CIR = Bombay, Baroda and Central India Railway

   EBR = Eastern Bengal Railway

   EIR = East Indian Railway

   GIPR = Great Indian Peninsula Railway

   GSIR, later SIR = Great Southern Railway of India, later South Indian Railway

   IVSR = Indus Valley State Railway

   MR = Madras Railway

   NWR = North Western (State) Railway—formed by the amalgamation under state auspices of the SP&DR and the PNSR

   PNSR = Punjab Northern State Railway

   SP&DR = Sind, Punjab and Delhi Railway

8. Archives and Libraries

   BM = The British Library's collection in the British Museum, London

   IOL&R = India Office Library and Records, London

   MSA = Maharashtrian State Archives, Bombay

   NAI = National Archives of India, New Delhi

   TNSA = Tamil Nadu State Archives, Madras

These abbreviations are part of a convention whereby they also serve to designate the originating level of a government record. Unless otherwise specified, a record grouping found in the IOL&R or the NAI was generated by the Government of India. For example, IOL&R, P/2750, PWD, RR Construction Progs., July 1886, no. 151 reads in full: Government of India, Public Works Department, Railway Construction Proceedings etc. with P/2750 being the modern IOL&R classification number and July 1886, no. 151 being the date and number of the proceeding. Provincial records found in the IOL&R are designated as such. An MSA designation means the record came from the Government of the Bombay Presidency and TNSA from the Government of the Madras Presidency.

9. Railway Letters

There are three record grouping in the IOL&R that are particularly valuable for the study of the earlier decades of railway construction. The abbreviation I use and the full title of each record series is:

Bengal RR Letters = L/PWD/3/40–113, Railway Letters and Enclosures from Bengal and India, 1845–79

Bombay RR Letters = L/PWD/3/245–285, Collections (Enclosures) to Railway Letters from Bombay, 1846–1879

Madras RR Letters = L/PWD/3/206–220, Collections (Enclosures) to Railway Letters from Madras, 1839–1879.

The letters are numbers consecutively within a calendar year. Some years had more than 300 letters. In most cases the enclosures provide the valuable material since the accompanying 'Letter' or despatch to the Secretary of State for India usually is brief and records the decision taken or advice given on the basis of the material in the enclosures. Since a letter often has a number of lengthy enclosures it is difficult and unwieldy to cite specific documents within enclosures. Usually I cite the letter and its date, e.g. Bengal RR Letters no. 5, dated 7 January 1859 or, when the exact date is unclear, just the letter number and years.

10. Journals and Publishers

EPW = Economic and Political Weekly

IESHR = Indian Economic and Social History Review

JAS = Journal of Asian Studies

JPS = Journal of Peasant Studies
MAS = Modern Asian Studies
CUP = Cambridge University Press
OUP = Oxford University Press
PUP = Princeton University Press

CHAPTER 1

Introduction

No railways operated in India in 1850. Twenty-five years later India had an extensive network of trunk lines. Fifty years later, in 1900, trains steamed through most parts of India along railways whose trunk and branch lines extended over 25,000 miles of track.[1] Railways, to paraphrase Theroux, had come to possess India and to make her hugeness graspable.[2] Nonetheless, the building of these railways remains a great and largely untold story of nineteenth-century India: a story of individual and collective effort, danger and quiet heroism, hardship and accomplishment. This book tells part of that story, the part that describes who conceived the railways of the Raj, who built them and how and under whose directions the builders worked. The process of railway construction in India in the last half of the nineteenth century, a process visualized as a grand assembly of many specific work processes, is here described and analysed.

This, then, is a book about reshaping and surmounting the natural world, the often rugged and sometimes treacherous terrain of India, to provide a permanent way within the capacities of steam locomotives; it is about the men, women and children who physically built the railways, the organization of their work and the materials, the tools and the machinery they used. Labour, tools, machinery, stone for ballast, wooden sleepers, lime, clay and mortar for bricks and brickwork, iron girders for bridges and the rails themselves all had to be obtained and transported to the worksites, so this book is also about a vast arena within which the constituent elements of the labour process, including capital, were assembled. The sources of some raw materials, such as wood, were located in India, as were low-level manufacturing processes such as brickmaking. The capital and most of the value-added products of heavy industrial

[1] Morris and Dudley, 'Railway Statistics', pp. 194–5: 23,627 route miles were open in 1900. Add double tracking and the running miles totalled 25,101 in 1900. Map 3 depicts the railways as they existed in 1861, 1881 and 1901.

[2] Paul Theroux, 'Introduction' to *Railways of the Raj* by Michael Satow and Ray Desmond (New York and London: New York University Press, 1980), pp. 6–7.

engineering needed for the railways, such as rails, bridge girders and work engines (stationary and locomotive), came from Britain. The nineteenth-century railways of the Raj were financed largely by British capital and built with British and Indian materials by Indian labour, supervised and directed by a few British colonial bureaucrats, contractors, engineers, foremen and skilled workmen.

A project of the magnitude and complexity represented by railway construction in nineteenth-century India required millions of workers with different kinds and levels of skills.[3] The work ranged from the complex abstractions of a consulting engineer in London designing a great bridge to the demanding simplicities of a coolie moving earth in the hills of Assam. The colonial bureaucracy, too, was heavily involved, from senior officials in the India Office to audit clerks and peons in mofussil India. The vast majority of the workers were manual labourers and overwhelmingly Indian, as were most of the gangers and petty contractors. Unskilled diggers and movers of earth and rock predominated, but many skilled workers were employed as well. Senior foremen, engineers, contractors, and bureaucrats provided most of the rest of the cast and they were usually British. This book, then, is also about the people who built the railways of the Raj and those who managed the construction process: who they were; where they came from; how they were mobilized, and something of the life they experienced at the worksites.

Above all, this is a book about the management of the construction process, that is management in the broad sense encompassing conception, the co-ordination of the process at a multiplicity of levels and the exercise of authority over workers engaged in the actual work of construction. Since management was a function dominated by the British—or at least the supervisory levels since Indians were present lower in the chain of command—this book is fundamentally about how the British got the railways of nineteenth-century India built. But the question 'how' raises the question of 'through whom', thus permitting additional questions to fit nicely within this organizing framework. For example, railway building was heavily labour intensive: it depended on an adequate supply of labour and the effective use of that labour, hence the extensive discussion of Indian labour in this study.

The focus on management does not deny the importance of many

[3] One authority states that it was 'the costliest construction project undertaken by any colonial power in any colony'. Daniel R. Headrick, *The Tentacles of Progress. Technology Transfer in the Age of Imperialism, 1850–1940* (New York: OUP, 1988), p. 53.

approaches that currently inform the social history of labour, but they must come later in the history of railway construction in India. A detailed, 'from below' socio-cultural history of Indian construction workers will require an understanding of what was above that gave shape and direction to the construction process. The prior task, the one undertaken here, is to establish a basic framework upon which subsequent work can be based.[4] The labour process is a complex phenomenon but, as Richard Price has correctly observed, 'it should not be forgotten that it is the search for authority that integrates the various aspects of the process'.[5] It was authority—management broadly conceived—that gave shape to the railway construction process, and is therefore a focus of this study.

Railway construction in India involved the directed effort of numerous hands involved in many and diverse activities. The work spanned continents and decades, but it all fed the same enterprise. One can choose to recognize the presence of particular activities within the assembly. For example, a brick-making compound in rural India was different from a heavy engineering factory turning out iron girders in industrial Britain, and both could operate without any connection to India's railways, yet both were important components of the construction process. Construction tools were made by previous labour processes, whether in British factories or by country artisans in India. Previously constructed boats and navigable waterways, roads and carts and, increasingly, operating railways provided the means by which construction materials got to the worksites.

[4] Modern published scholarship on India's nineteenth-century railway history has focused largely on government policy, financing, administrative structures, rolling stock and, more recently, on the economic impact of the operating railways. Useful as a summary of recent thinking is John M. Hurd, 'Railways', in *CEHI*, pp. 737–61. The need for a modern study of railway construction is supported by the fact that even in the mid-1990s the most useful book on nineteenth-century Indian railway construction remains Captain Edward Davidson, *The Railways of India. With an account of their rise, progress and construction written with the aid of the the records of the India Office* (London: E. and F.N. Spon, 1868). It was written in the early phase of construction by a Royal Engineer who, in his capacity as Deputy Consulting Engineer for Railways to the Government of India, had a close connection with the construction process. Modern writing on Indian railway construction is restricted to a number of helpful chapters in semi-popular treatments such as P.S.A. Berridge, *Couplings to the Khyber. The Story of the North Western Railway* (New York: Augustus M. Kelley, 1969).

[5] Richard Price, 'Structures of Subordination in Nineteenth-Century British Industry' in *The Power of the Past. Essays for Eric Hobsbawm*, ed. Pat Thane, Geoffrey Crossick and Roderick Floud (Cambridge: CUP, 1984), p. 119.

But it was the assembly, direction and use of these and other components that made railway building in India possible. If the British provided capital and technical knowledge they also integrated, supervised and directed; they managed in the broad sense the grand assembly of activities I label the construction process. The British were, by and large, the managers and the technical experts, the mental labourers in the construction process.

My analysis of the management of the railway construction process is informed by the fact that British capital sponsored and directed the building of India's railways. Some £150 million of British capital was invested in India's nineteenth-century railways, the single largest investment within the nineteenth-century British empire.[6] Bartle Frere, Governor of Bombay Presidency, did not doubt that railway construction was a project of capitalism. His speech at the opening of the Bhore Ghat incline in 1862 was an extended paean to the beneficial effects of the expenditure of British capital on railways works. What he called the 'Railway Period' had freed individuals to sell their labour power in a market regulated by 'the natural laws of supply and demand'.[7] 'For the first time in history the Indian Cooly finds that he has in his power of labour, a valuable possession. . . .'[8] A few years later Lord Napier, Governor of Madras, also did not doubt that the social relations of production at the worksites were those of capitalism when he said in 1868 of the contractors who built the Chitravati bridge on the northwest line of the Madras Railway: 'They temper the relations of capital and labour with an ingredient of kindness which endears them to the people.'[9] British

[6] The amount is a matter of debate. B.R. Tomlinson, *The Political Economy of the Raj, 1914–47: The Economics of Decolonization in India* (London: CUP, 1979), p. 4 claims that some £ 200 million was invested before 1914. Daniel Thorner, *Investment in Empire* (Philadelphia: University of Pennsylvania Press, 1950), p. viii asserts that the British capital which went into Indian railways 'formed the largest single unit of international investment in the nineteenth century'.

[7] *Railway Report, 1862–3*, p. 27. The most recent study of Frere's years in Bombay (1862–7) is Rekha Ranade, *Sir Bartle Frere and His Times* (New Delhi: Mittal Publications, 1990).

[8] *Railway Report, 1862–3*, p. 27.

[9] *Madras Times*, 10 January 1868. The occasion was the opening of this major bridge which was marked by the standard celebration and speeches. Napier went on to say that the contractors had 'attained great influence and authority in the country, with a command and confidence in the labour market, which entitles them to undertake and complete their works with unusual promptness and punctuality—a matter of the highest moment if we remember the important financial and social

investors and the Government of India invested in the railways. The means of production (the land, the materials and the tools) belonged to a capital that came to mean both private and state capital. Men with capital (or state capital) or their factotums directed the work of construction and owned or managed the product, the operating railways. The men, women and children who built the railways had only one connection with the railway companies: the sale of their labour power. They became wage earners whose connection with the railways was the money they earned as construction workers. However, and it is an important caveat, the construction workers also often had extra-economic relationships with the petty contractors or headmen who usually directly commanded their labour.

The widespread transformation of labour power into a commodity that the worker must sell to survive is a feature of the transition to capitalism—an important marker of the advance of capitalist relationships although not, of itself, the defining characteristic of the transition. In the teleology of both marxism and capitalism it is both a necessary and a liberating development, however fraught with unpleasant consequences for the worker the process may be in the short run. Certainly, British advocates of Indian railway development had no reservations about the value of the wage connection. Frere did not, and although others wrote in a similar vein he, in the ripeness of his prose, the extent of his vision, and the sweeping confidence of his assertions, captured the essential vision of the Victorian advocates of railways for India.[10] If the British-sponsored millennium was on its way, the railways, the Victorians believed, would hasten its arrival in India. English capital in large amounts was being expended on Indian railways and setting in motion significant social and economic changes. Railway construction work was giving Indian villagers an opportunity to sell their labour power, thus freeing them from the thralldom of a serf-like existence.[11] In 1860 one observer stated that in the previous five years £14,000,000 had been spent on railway construction labour in India while in 1868 the contractor Charles Henfrey, an associate of Thomas Brassey, estimated that of the £75 to £80 millions expended on Indian railways £40 to £50 millions had passed 'into the hands of the

---

interests which are involved in the development of the railway system'. We are likely to be sceptical of Napier's reference to the kindness of the contractors who, then as now, usually were not generous to those they employed.

[10] Nor were they all British. Read the equally optimistic vision presented in Framjee Vicajee, *Political and Social Effects of Railways in India* (London, 1875).

[11] *Railway Report, 1862–3*, p. 27.

working classes'.[12] Railway construction in nineteenth-century India was a capitalist enterprise, a fact not changed when the colonial government began to build and own railways.

But how in this huge enterprise of Victorian capitalism were workers mobilized and used? How did labour get to the work-sites in sufficient quantity and how was it directed and organized once there? To what extent and in what ways did British engineers, contractors, and overseers who had already accepted the dictates of industrial capitalism introduce the working methods of capitalism to the tasks of railway construction in India? How, from the perspective of management, were particular work processes and the grand assembly of those processes organized and directed?

The assertion that the labour process and changes thereto are important, has a distinguished lineage. Marx certainly thought so—as Adam Smith did before him—and devoted much of volume I of Capital to a description of the nature, direction, and consequences of the capitalist labour process. Indeed, Marx wrote on the subject with a power almost amounting to closure, such that for generations few within the Marxist tradition paid much attention to the labour process. It took a work of insight and powerful elegance to reassert the centrality of the labour process to the study and critique of capitalism. That work was Harry Braverman's *Labor and Monopoly Capital*.[13]

Braverman's work provided the stimulus for this study. It is in the following passage from Braverman that I find the dilemma that the contractors and engineers faced in India; a dilemma that they continually sought to overcome and one that highlights the theoretical significance of the discussion of the conception, co-ordination and control of the labour process in this study. Given the indeterminacy of labour power, how is the extraction of surplus-value ensured?[14]

But if the capitalist builds upon this distinctive quality and potential of human labor power, it is also this quality, by its very indeterminacy, which places before him his greatest challenge and problem. The coin of labor has its obverse side: in

[12] Berkley, *MPICE*, 19 (1859–60), 615. Arthur Helps, *Life and Labours of Mr. Brassey 1805–70* (London: Bell and Daldy, 1872), p. 45.

[13] Harry Braverman, *Labor and Monopoly Capital. The Degradation of Work in the Twentieth Century* (New York and London: Monthly Review Press, 1974).

[14] Michael Burawoy, *The Politics of Production* (London: Verso, 1985), ch. 1, esp. pp. 30–5 refers to this as the problem of 'obscuring and securing surplus value'. Burawoy, although a major critic of Braverman, is also generous in his praise of Braverman's accomplishments.

purchasing labor power that can do much, he is at the same time purchasing an undefined quality and quantity. What he buys is infinite in *potential*, but in its *realization* it is limited by the subjective state of the workers, by their previous history, by the general social conditions under which they work as well as the particular conditions of the enterprise, and by the technical setting of their labor. The work actually performed will be affected by these and many other factors, including the organization of the process and the forms of supervision over it, if any.[15]

How, in concrete terms, was the labour power of Indians used to build railways? What were the particular conditions of the enterprise, the technical setting, the organization of the process and the forms of supervision? To repeat, how did the British get the railways built?

One answer in Indian railway construction frequently was, as it has often been in the history of capitalism, for capital to delegate the tasks of production to contractors and sub-contractors.[16] Often in the Indian case the salaried engineers of the railway companies (private and state) functioned like contractors since they got the work done through sub-contractors. Contract work continues to be a common practice within capitalism and it is not a form of management peculiar to early capitalism or representative of a need of capitalists to avoid the tasks of management. It is an effective response to certain conditions of production. It is, in fact, a response that is becoming more common in the conditions that prevail in the capitalist world economy at the end of the twentieth century.[17] Contractors and sub-contractors working under the technical and managerial direction of professional engineers appear in a prominent role in this book.

Marx distinguished between two varieties of the capitalist labour process that he labelled the formal and the real subsumption of labour under capital. Formal subsumption was 'the general form of every capitalist process of production' that, moreover, could persist as 'a particular

[15] Braverman, pp. 56–7. Emphasis in the original.

[16] See Sidney Pollard, *The Genesis of Modern Management* (London: Edward Arnold, 1965), esp. pp. 38–47. Also on this and other points read the stimulating Charles Sabel and Jonathan Zeitlin, 'Historical Alternatives to Mass Production: Politics, Markets and Technology in Nineteenth-Century Industrialization', *Past and Present*, 108 (August, 1985), pp. 133–76.

[17] See the concluding comments in Maxine Berg, Pat Hudson and Michael Sonenscher, 'Manufacture in Town and Country before the Factory', in *Manufacture in town and country before the factory*, ed. Berg, Hudson and Sonenscher (Cambridge: CUP, 1983), pp. 30–2.

form alongside the specifically capitalist mode of production in its developed form'.[18] The mode of production itself did not change and the labour process, although subordinated to capital, changed little with respect to work practices and technology.[19]

> Within the production process, however, as we have already shown, two developments emerge: (1) an economic relationship of supremacy and subordination, since the consumption of labour-power by the capitalist is naturally supervised and directed by him; (2) labour becomes far more continuous and intensive, and the conditions of labour are employed far more economically, since every effort is made to ensure that no more (or rather even less) socially necessary time is consumed in making the product. . . . But the more capitalist production sticks fast in this formal relationship, the less the relationship itself will evolve, since for the most part it is based on small capitalists who differ only slightly from the workers in their education and their activities.[20]

Marx then established a striking contrast between formal subsumption and the real subsumption of labour that occurred, he believed, with the transition to the developed capitalist mode of production. The labour process itself was transformed and became, in his view, the real capitalist labour process. Increased co-operation, the greater division of labour, 'the use of machinery, and in general the transformation of production by the conscious use of the sciences . . . technology etc.' enormously increased production, and with it the extraction of relative surplus value.[21] Labour became more thoroughly subordinated to the dictates of capital that could use machinery 'to control labour through the production process itself'.[22]

The concept of formal subsumption is valuable and captures much of what occurred in the building of India's railways. Pre-existing work processes effectively served the cause of construction. The petty Indian contractors who became the main instruments for the mobilization and utilization of unskilled construction labour fitted well Marx's characterization of petty capitalists. The concept of a change from formal to real subsumption is also valuable. Aspects of railway construction in India did, and did so increasingly, involve the application of science and technology, the use of machinery, and the increased division of machine-paced labour.

[18] Marx, *Capital*, vol. 1, trans Ben Fowkes (New York: Vintage Books, 1977), p. 1019.

[19] Ibid., pp. 1020–1.

[20] Ibid., pp. 1026–7.

[21] Ibid., p. 1024.

[22] Paul Thompson, *The Nature of Work* (Houndsmill: Macmillan, 1983), p. 124.

However, Marx's two dimensions of the labour process could and did exist at the same work-site. Indeed, teleology notwithstanding, it is more helpful to view formal and real subsumption as dimensions of the same phenomenon (the labour process) with no necessary, unidirectional movement from the former to the latter.

Recent scholarship has shown how various means of production and relations in production, i.e. different forms of the labour process, can be made to serve the interest of capital.[23] Burawoy discards the notion of a distinctive capitalist labour process in favour of the view that the same labour process (for Burawoy, 'relations in production') can be found in different modes of production.[24] I accept Burawoy's position and find it exemplified in the building of India's railways. One must view the construction process in a way that recognizes the presence of a diversity of managerial and labour practices—a diversity magnified by the fact that nineteenth-century colonial India was a world in complex and uneven transition from a previous mode of production to the more encompassing presence of the capitalist mode of production, a transition speeded up by railway construction and its offspring, the operating railways of India. Much of the work, especially earthwork, was characterized by formal subsumption in the 1850s and in the 1890s, but it is also correct to say that by the 1890s earthworking was, from the managerial perspective, better controlled and more productive. One should not underestimate the productive returns to management represented by longer, more intensive, more continuous, and especially more orderly work. Other aspects of the assembly, such as brick-making, lime-making or the erection of iron bridges, came increasingly to resemble Marx's category of real subsumption.

Those who managed the construction of India's railways sought to organize the process in the ways they believed (and they could and did misread situations) most likely to get the job done expeditiously and/or economically. Economically usually meant within some previously deter-

[23] That interest, of course, is the realization of surplus value. The exploitation of labour by capital which is inherent in the valorization process forms the central nexus of the relations of production which define the capitalist mode of production.

[24] Burawoy, p. 14 *et passim*. Thus, for Burawoy, the distinguishing features of the capitalist mode of production are to be found in the relations of production. Since I am concerned in this study with production, with how the railways got built, I only indirectly touch upon the large and fascinating topic of the extent to which railway building helped to intensify the presence of capitalist relations of production in nineteenth-century India. To what extent, for example, did railway construction contribute to the proletarianization of rural India?

mined cost limit, e.g. for a major contractor it would be his tendered amount for the contract minus the amount of profit he expected to make, while expeditiously meant adherence to a dictated or expected schedule for completion.[25] Control of the labour process was sought through whatever methods were believed to be most efficacious in particular contexts and times. The most effective managers were often those who recognized the value of flexibility. Control could involve the use of extra-economic methods. Control also could be exercised through the wage structure (piece-work versus time-determined wages, for example), through the methods and timing of wage payments and through technical innovations. Workers needed their wage to maintain themselves and their families. The wage was the main weapon of employer control. Most worker resistance involved the security and the amount of wage payments. A task of this study, therefore, is to capture, in the contexts of railway construction in India, the diversity of managerial methods through which the British got the railways built. This study also will seek to show that managerial control of the construction process did become more effective as the decades passed, with effectiveness defined as the ability of the British to get railways built more expeditiously and/or economically.

Effectiveness did not mean that managerial control over specific labour processes was necessarily tightened. Expedition and/or economy could often be achieved best, especially in earthworking at the level of the gang, with tools and patterns of work organization that preceded India's railway period, while simultaneously expanding the co-ordinated use of many gangs: a development that could be labelled extensive as opposed to intensive management. The juxtaposition of different kinds of labour processes was expressed well by a twentieth-century engineer in another type of construction work in India, dam building. In the 1930s, despite the use of modern machinery, some four-fifths of the Mettur Dam in South India was built by hand, leading the engineer to observe: 'In no other country, perhaps, could such methods have been employed economically or such a high rate of outturn maintained.'[26]

It also must be emphasized that the hierarchy of co-ordination and control was not a monolith. The Government of India usually wanted

[25] Economically did not mean cheaply. India's nineteenth-century railways rarely were built cheaply. See chapter two. However, cost limits were lowered and more stringently applied as the half century progressed so, in that sense, the projects became cheaper.

[26] C.G. Barber, *History of the Cauvery-Mettur Project* (Madras: Superintendent Government Press, 1940), preface.

construction to proceed both economically and expeditiously, but at lower levels there could be different emphases. Different levels of managers on the same project could have different interests. A chief engineer under instructions from the government or from a major contractor with a deadline might have wanted construction to proceed expeditiously and economically. A sub-contractor for a segment of the same line was most likely interested in maximizing his profits, so his strategy might be to pursue low cost construction that might be less attainable if the high labour inputs needed to construct quickly pushed up wage rates. A ganger with extra-economic relationships with his or her gang had yet other concerns. At the other extreme a line needed urgently for military purposes might be built quickly despite cost. To all of this the crucial group of professional engineers often brought another goal, the goal of an increasingly institutionalized profession to build things well; to build railways according to the increasingly rigorous standards of their profession. The many accounts by British engineers of their Indian railway bridges, inclines or even entire lines in the engineering journals were not just for the information and education of their fellow professionals; the accounts were affirmations of personal and professional accomplishment.

British contractors and engineers attempted to apply the techniques, technology, and labour practices of British railway building to India and in the process they learned to adapt to and learn from India and Indians. A measure of what they learned is the fact that the British subsequently applied techniques acquired in India and Indian labourers to railway construction in Africa and South East Asia.[27] Railway construction from the 1850s through the 1890s, viewed as the development of certain productive processes, took place within a natural and social world about which the British, at first, knew little. In terms of the natural world, the effects of Indian river conditions on railway bridges were largely unknown. In terms of the social world, the British had to deal with Indians who had long-established forms of work that either had to be accepted, modified or replaced in the search for ways to build India's railways. Indians in turn either accepted or opposed the modifications or replacements. Technological transplantation, innovation and adaptation was present, but so was the continued use of long-standing Indian means of production that, as experience often demonstrated, could be more efficacious under Indian conditions than more sophisticated tools and techniques from Britain. An evolutionary development of tools and

[27] See Headricks, pp. 49–96, for a brief survey.

machines characterized the changes to the technical setting, but what evolved, often as not, were tools or machines whose prototypes were Indian.

Beyond technology there was always the organization of large-scale work. Obtaining, retaining, and using labour was an imperative task for management. The organization of large bodies of workers was a central feature of railway construction in India, as it was elsewhere. Large-scale gang labour using simple tools was the major form of work. Manuals for civil engineers in India devoted many pages to the problems associated with finding, organizing, retaining and paying large bodies of workers: managerial competence was as important to the railway engineer in India as technical knowhow. The problem was not unique to India, though in India the scale was greater. Most engineers in India would have agreed with George Stephenson (of 'Rocket' fame) who reportedly said: ' "I can engineer matter very well, but my great difficulty is in engineering men".'[28]

What gives a certain unity to the period 1850 to 1900 is the evolutionary development of the construction process. Also, the 1890s saw the completion of the basic system of trunk lines in India with the construction of such lines as the Bengal–Nagpur Railway, the East Coast State Railway and the Bengal–Assam line, thus also giving to 1850–1900 the unity of major trunk line construction. The direction of the development was clear. From the perspective of management, railway construction as a total process became seen as an increasingly more effective and better-managed process. The construction process became routinized at all levels. Work became more continuous and more orderly. Construction proceeded more expeditiously and/or more economically and, likely, what was built was usually built better—certainly bridges stopped falling down at their initial, alarming rate. It also may be that the physical conditions of work improved for the Indian workers and for British engineers, contractors' agents and overseers. It was more efficient for capital—railways were built with fewer interruptions—if, for example, Indian labourers were housed under better and more sanitary conditions. A healthy workforce was more productive and more stable. Even in the area of health there was an evolutionary development in the understanding of diseases like cholera and malaria and in the best methods to combat them in the context of railway construction. Those who managed the construction process slowly came to know much better how railways

[28] Robert Stephenson quoting his father, as reported in *The Engineer* (25 April 1856), p. 233.

should be built in India. A process only poorly understood in the 1850s and 1860s—whose first model was railway construction in Britain—had by the 1890s become familiar, routinized to a considerable extent, and better adapted to the natural and social conditions of India. Macgeorge captures well the transition:

> Rapidity, in the early construction of railways in a new country situated as India was forty years ago, was hardly to be expected. Practically nothing in the way of manufactured materials or working plant was purchasable in the country; and the engineers and contractors had to manufacture, prepare or collect—often from great distances—everything required, such as bricks, stone, lime, timber, iron-work, or plant of all kinds. At the commencement of operations, and for some time subsequently, no trained subordinate staff existed, and inspectors imported from England—then a vastly more distant country than at present—were almost useless until they had acquired some knowledge of the language, and had learned and unlearned much new and old experience; and lastly, skilled labour of almost every description did not exist, until in the course of time it could be trained and educated, and even ordinary unskilled labour was difficult to secure, until full confidence in new employers had been gained.[29]

The evolution of the construction process did not end in the 1890s but the period under study marked, for the British, the transition from the difficult, mistake-ridden, experimental situation of the first two decades to the settled confidence of the 1890s and beyond. For example, one important feature of this transition was the steady growth of a body of Indian workers and the petty contractors who hired them. Both groups turned to construction (not solely to railway construction) as their main form of employment. Moving from project to project these workers—heavy duty navvies and skilled workmen—made the acquisition of labour for particular projects easier. Moreover, because these workers acquired and transmitted railway building skills among themselves, they expedited the work of construction.

Indian workers may well have viewed many of these developments in a different and less positive light. There was resistance and collective action, including strike action, among the workers. The British often viewed resistance to technical innovation as blind prejudice, while strike action was something to be crushed if possible. A later chapter deals briefly with worker resistance, for it was a part of the labour process. 'The labour process . . . is above all else a social process in which the technical characteristics of a particular work environment shape and condition the

[29] G.W.Macgeorge, *Ways and Works in India* (Westminster: Archibald Constable and Company, 1894), p. 310.

forms of struggle for authority and control. It is in the continual search from both sides for a better bargain that the dynamics of the labour process in labour's history can be seen to lie.'[30] The engineers and contractors came to recognize that to get a project finished quickly and/or economically they had to control labour in ways that were situationally expedient even if, finally, they held the whip-hand. Labour in turn resisted, modified or accepted the managerial initiatives. The outcome was situation specific, depending, among other things, on the respective strength of capital and labour as embodied in real workers and those placed over them at particular work-sites.

The construction workers were never passive participants in the labour process; they accepted, exited, or gave voice to protest.[31] One needs to know how the British dealt with resistance and how the resistance affected the evolution of the construction process. However, this is not a history of worker agency; it is a history of how the British got the railways built. That the workers had agency I accept, but, except tangentially, this book does not seek to explore that agency. Moreover, when that agency is explored in later studies, nineteenth-century construction workers will never be known first hand and from below; we will only know them from above and refracted through British, colonial eyes. Nineteenth-century construction workers will always be mute: people spoken of, who left no first-hand accounts of their own.[32]

[30] Richard Price, 'The Labour Process and Labour History', *Social History*, 8:1 (January 1983), p. 62.

[31] I agree with Marx that the historical direction, though not unilinear, is towards the greater control of labour by capital (though not necessarily in the ways he envisioned) and that the dynamic, the impetus, to do so lays within the logic of capitalism. At least historically (the telos, hopefully, will be different) the workers are usually in the position of reacting to managerial initiatives. I therefore cannot embrace the 'terrain of compromise' argument to the extent fashionable among some social historians. David Washbrook, 'Progress and Problems: South Asian Social and Economic History, c. 1720–1860', *MAS*, 22:1 (1988), p.91 is on the mark where South Asia is concerned: 'Political control over the forces of market competition and increasing dominance over labour and its processes of reproduction gave capital a comfortable history in colonial South Asia'.

[32] The sources that exist, and they will, perforce, be primarily official sources since Indians did not write much about labour matters in the nineteenth century (i.e. before an organized labour movement brought into play literate, middle-class Indians with an interest in labour matters) let alone about construction work and workers (the construction workers themselves were overwhelmingly illiterate), will have to be interrogated creatively, with the silences being read as much as the texts. One could look to the stimulating Dipesh Chakrabarty, *Rethinking Working-Class*

The ensuing chapters put historical flesh to the concerns sketched in this introduction. Chapter two attempts to contextualize certain central elements of the railway construction process. The patterns of ownership, co-ordination and control of the process are identified. The genesis and nature of the special role of the Government of India in the supervision of the construction is described. The start-up conditions of the 1850s are established as a base from which to view subsequent developments. The vicissitudes of subsequent construction activity are surveyed. An attempt, summarized in Table 1, is made to estimate how many miles of track were actually under construction each year between 1850 and 1900, since the intensity of the construction activity determined the size of the construction workforce in any year. An additional estimate, whose sources and assumptions are detailed in the appendix, gives the average number of Indians employed to build one mile of railway that, when wedded to the estimates of miles under construction, generates (Table 2) estimates of the numbers of railway construction workers employed annually from 1859 to 1900. Chapter three talks about the organization and direction of the construction of specific lines or sections thereof; it is about the varieties of ways in which the British managed the labour process at the point of production. Chapter four focuses primarily on those whose labour power physically built the railways: the Indian carpenters, riveters, masons, brick-makers, divers, riggers and, in their tens of thousands, the earthwork coolies. How were the large numbers of workers recruited? From where? And what can one say about the social and occupational composition of the workforce? How, in short, did the railway builders obtain workmen in the numbers and with the skills needed? Chapter five turns to the questions of work and working conditions. Work processes are described, as are the conditions of work and life at the construction sites—harsh conditions usually endured that sometimes bred anger and collective resistance. Chapter six examines some examples of resistance among the Indian segment of the construction workforce from the perspective of how the British dealt with what they considered to be an obstacle. The concluding chapter summarizes the main findings and returns briefly to the issues raised in this introduction.

History. Bengal 1890–1940 (Princeton: PUP, 1989) for a possible way forward, although I cannot embrace fully his 'cultural Marxism'.

CHAPTER 2

Getting Started and Subsequent Progress

A feature of advanced capitalism as manifested in large-scale enterprises involving substantial amounts of capital—as the railways of India most certainly were—is the separation of ownership and control. This was so with Indian railways, where most private shareholders were uninvolved in the direction of the railway companies. However, from the start the Government of India was involved in the supervision and control of the construction process and, by the late 1860s, also began to build and operate a state railway system. As one legal scholar observed, the 'dominant characteristic' of Indian railways, present from their very beginnings, was the interest the Government of India held 'as guarantor, or partner, or proprietor' in the entire system.[1] The beginnings of what became an increasingly complicated system of ownership and control need to be described, and the consequences of that system for the management and conduct of the construction process explored.

Years of promotion and debate preceded the building of railways in India. The advocates, the polemicists, the promoters, the visionaries, the would-be speculators—often one and the same—produced a steady barrage of printed material touting their particular schemes and extolling the benefits that Indian railways in general or specific lines in particular would bring to investors, to the Indo–British commercial connection, to the security of British rule and to the people of India. Promoters scorned the schemes of their rivals and lobbied intensively in public and private for the right to build particular lines. Few critics or cautionary voices were to be heard: railways were clearly a good thing for India.[2] Critics of various

[1] Henry Edward Trevor, *The Law Relating to Railways in British India* (London: 1891), p. 1.

[2] The campaign for railway construction can be followed in Daniel Thorner, *Investment in Empire. British Railway and Steam Shipping Enterprise in India 1825–1849* (Philadelphia: University of Pennsylvania Press, 1950), pp. 44–118; W.J. Macpherson, 'Investment in Indian Railways, 1845–1875', *Economic History Review*, 2nd. series, 8:2 (1955), pp. 177–86.

sorts and with varied targets did appear in later years as the costs and difficulties of Indian railway building became known. Sir Arthur Cotton asserted strongly that the money spent on railways could have been more usefully spent on irrigation projects and navigable canals—a position subsequently endorsed by Naoroji and R.C. Dutt. Other Indian critics pointed out the hidden and not so hidden costs of railway construction but, as Bipan Chandra notes, they did not question the need for railways and chose to concentrate on the pace at which railway construction should proceed.[3] But that was in the future; at mid-century the railway promoters were unchallenged.

Active promotion notwithstanding, British capitalists were reluctant to risk their money on Indian railway ventures and the Directors of the East India Company had no wish to build railways as a state-financed effort; and, at least initially, the Directors had little enthusiasm for guaranteeing to private capital an assured return on investment in Indian railways. Thus, despite the extensive lobbying and public promotion, the 1840s drew almost to a close without any firm decision to begin construction. The sticking-point was the question of the guarantee.[4]

Finally, in March 1849—the same month in which major British expansion by force of arms on the Indian subcontinent was completed with the annexation of Punjab—the East India Company agreed on terms with the Great Indian Peninsula Railway (hereafter GIPR) and the East Indian Railway (hereafter EIR), whereby the two companies would build and operate their respective lines with a guaranteed five per cent return on their stockholders' investment, assured by the revenues of the Government of India. The preliminary agreement of March was followed by the extensive, formal contracts of 17 August 1849.[5] The legal groundwork had been completed; embankments, cuttings, bridges, tunnels and the permanent way soon followed as the first phase of railway construction in India commenced as 'private enterprise at public risk'.[6] The investment came almost entirely from Britain but the risk was born by Indians whose taxes paid the difference between the five per cent guarantee and the lower rate of profit the guaranteed companies consistently earned throughout

[3] Bipan Chandra, *The Rise and Growth of Economic Nationalism in India. Economic Policies of Indian National Leadership, 1880–1905* (New Delhi: People's Publishing House, 1966), p. 177. He surveys the positions of various Indian critics on pp. 177–216.

[4] Thorner, especially pp. 119–67; Macpherson, pp. 180–6.

[5] Thorner, pp. 119–82 has a fine summary of the struggle to obtain the guarantee and the resultant contracts.

[6] Thorner, p. 168.

the nineteenth century. With British investment in Indian railways totalling some £150 million in the nineteenth century, of which some £95 million were invested by 1875, the revenues of the Government of India were tapped for some £50 million to meet the guarantee.[7] Guaranteed returns to capital, however, were not unique to Indian railways in the nineteenth century. Davis and Huttenback note that most railways built in the nineteenth century outside the United States received some kind of government guarantee or subsidy.[8]

The contracts with the EIR, the GIPR and subsequent private companies are central documents in the history of the Indian railways. However, the student of railway construction needs only to focus on a few aspects of these formidable documents, namely the clauses that specified the nature and degree of control and supervision government obtained as its *quid pro quo* for the guarantee: the guarantee that Lord Canning in 1858 was to refer to as 'a species of equivalent tendered by Government in purchase of that direct control which it is so important to maintain over these great undertakings'.[9] Clause 2 of the EIR contract gave the East India Company—as subsequent or inherited contracts gave the Crown—the right to determine the route, direction and length of the lines. Clause 10 asserted the subjugation 'in all things' of the EIR, its officers and its employees 'to the superintendence and control of the East India Company'.[10]

[7] The exact amounts are the subject of some disagreement. See Daniel Thorner, 'Great Britain and the Development of India's Railways', *Journal of Economic History*, 11: 4 (fall 1951), pp. 391–2; Leland H. Jenks, *The Migration of British Capital to 1875* (1927; reprint edn, London: Nelson, 1971), p. 225; A.K. Banerji, *Aspects of Indo-British Economic Relations 1858–1898* (Bombay: OUP, 1982), pp. 50–76. Banerji attempts to put the issue on the firmest quantitative footing but he too leaves many uncertainties.

[8] L.E. Davis and R.A. Huttenback, *Mammon and the Pursuit of Empire* (abridged edn; Cambridge: CUP, 1988), p.283, note 24.

[9] Canning, Governor General's Railway Despatch no. 2 of 29 November 1858, quoted in W. Eric Gustafason, 'The Gift of an Elephant? The Indian Guaranteed Railways, 1845–1870', paper presented at the Annual Meeting of the Association for Asian Studies, March 1971.

[10] *Parliamentary Papers* (Commons), 1859, sess. 1, cmnd. 259, 19 April 1859: 'Copies of all Contracts and Agreements entered into by the East India Company, or by the present Council for India, with any Company formed for making Railways, Public Roads, Canals, Works for Irrigation, or other Public Works in India', pp. 3–6. The particular contract quoted here is with the EIR, but in most matters the contracts had similar or identical wording. Later returns to Parliament provide

Further material in Clause 10 gave the East India Company the right of access to virtually all the accounts, proceedings, minutes, papers, etc. of the Railway Company and to appoint, ex officio, a member of the Railway Company's Board with 'a right of veto in all proceedings whatsoever, at Boards of the said Directors'.[11] Other clauses listed additional powers. In short, the contracts gave the Government of India extensive powers over most aspects of railway development, construction and operation in the Indian subcontinent.

The Government of India exercised extensively its rights of supervision and direction. In London, in Calcutta, in the capitals of the presidencies and the provinces and on inspection tours, the officials of government involved themselves in what was often the most minute details of railway building. In the event, as a number of historians have demonstrated, other clauses in the railway contracts and other features of the situation worked directly against one of the central goals of government supervision, namely economy of construction.[12] At an average cost of £18,000 per mile the railways built in India in the 1850s and 1860s were not cheap and were, for some, responsible for India's finances, a 'gift of an elephant'.[13] The Guaranteed Companies, moreover, had no land assembly costs and small legal costs because government provided the right of way.[14] Ineffectual cost control notwithstanding, government supervision and direction made the officials and their procedures a central feature of the entire process of construction from deciding where and what to build to the declaration that a particular line was fit to be opened for traffic. The same government supervision, direction and record keeping that formed a part of the process made this book possible. It is unlikely that

---

collections of contracts up to 1871 and 1882 respectively. They, along with the annual Railway Reports and some other financial returns, are conveniently collected in the IOL&R series, L/AG/46/42/14–17.

[11] *PP* (Commons), 1859, sess. I, cmnd. 259, p. 6. The practice soon evolved of having one individual serve as the 'Government Director' on all the individual boards of the various guaranteed companies. Sir Juland Danvers (1826–1902) served in this capacity from 1861 to 1892.

[12] Thorner, *Investment*, pp. 173–5; Banerji, p. 54.

[13] Jenks, p. 222. This phrase was used by Sir Charles Trevelyan, Governor of Madras, in 1859 to describe the railway system that was exhausting 'our resources' and 'eating us out of house and home'. See Gustafson, p. 1.

[14] Such costs formed a substantial percentage of the £42,486 per mile average cost of railway construction in Britain up to 1884. See Philip S. Bagwell, *The Transport Revolution from 1770* (London: B.T. Batsford Ltd., 1974), pp. 99–102. Nonethe-

the voluminous bodies of sources on which this study is based would have been created at all, let alone survived, if it had not been for the detailed involvement of the Government of India.

Within two decades the elephant came to be seen as excessively expensive. So, in 1869, the Government of India embarked upon an extension of the railways through a system of State construction and State administration. State capitalism came early to India. This departure sparked considerable controversy in India and Britain. At the same time, the old, private, guaranteed companies continued to extend and operate their lines, although many eventually came under State ownership and State management. A variant pattern was State ownership and private company management. The mid-1880s saw a return to the construction and working of lines through the agency of private companies whose contracts, however, were different from those in the 1850s and 1860s, and whose guarantees were reduced to 4 per cent or lower. Thus, by the turn of the century, the railways in India came to be classified for administrative purposes of government into the following ten categories based on ownership and management criteria: (1) State-owned lines worked by private companies; (2) State-owned and State-worked lines; (3) Lines owned and worked by private companies guaranteed under old contracts; (4) Lines owned and worked by private companies guaranteed under new contracts; (5) District Board Lines (short, local lines within a district and paid for by local cesses); (6) Assisted Companies' Lines (government assistance of various sorts but no guarantee); (7) Princely State Lines worked by private companies but owned by the Princely State; (8) Princely State Lines worked by State Railway Agency, i.e. as part of the State system but owned by the Princely State; (9) Lines owned and worked by Princely States; (10) Lines in foreign territory (e.g. in French or Portuguese India).[15] Nonetheless, through all the changes government supervision of the construction process continued.

The changing modalities of financing, agency of construction and management probably made a difference to the railway map of India. The government, for example, was more likely to undertake projects for military, political or even humanitarian (e.g. famine relief) reasons. Even

less, the per mile cost of railway building in India was not disproportionately high, at least not if the comparison is made with Europe.

[15] The categories are those listed in N. Sanyal, *The Development of Indian Railways* (Calcutta, 1930), p. 189, in which the shifting modalities of financing and management can be followed most easily. Further changes took place in the twentieth century.

with a guarantee, private companies expected to build lines that would eventually generate a profit. And it was the Government of India, for reasons of anticipated economy in the face of the costs of the early guarantee system, that embarked on meter gauge construction in the late 1860s and thus introduced a break of gauge and an unending controversy over the wisdom of that policy.[16] But these were differences that had few consequences for the process of construction. Perhaps the meter gauge required the use of less labour. Perhaps construction by the State generated some additional career opportunities for the engineering bureaucrats of the Public Works Department. Perhaps private companies were more inclined to build their railways through contractors rather than through the agency of their own engineers. But these are arguable propositions with many nuances and qualifications. The more defensible generalization is that, on the whole, the actual construction of a line was little affected by the presence of state versus private modalities because the state was actively present in both situations. The general trend from the early 1860s onwards was for the Government of India to exercise more central control over all aspects of railway activity in India.[17] Government involvement was one of the main unifiers of the construction process. The detailed involvement of government was one of the distinctive features of railway building and railway operations in India. The government, of course, was a colonial government that provided few opportunities for the ruled to influence the formulation and execution of railway policy.

The construction process came to have three levels in so far as supervision and direction were concerned: the level located in Britain; the level located in India at the imperial capital of Calcutta and at various

[16] Only one, relatively short, branch line was built with a gauge less than 5′ 6″ prior to State construction. That line, the Nalhati Railway from Nalhati to Azimganj, was built with a 4-foot gauge; it remained thereafter a solitary anomaly. At the turn of the century (1903), 14,477 miles were of the 5′ 6″ gauge, 11,421 of meter, 796 of 2′ 6″, and 262 miles of the 2-foot gauge. The gauge controversy can be most easily followed in four articles in *MPICE*: Thornton, 35 (1873), pp. 214–535; Waring, 97 (1889), 106–94; Upcott, 164 (1906), 196–327; Royal-Dawson, 213 (1922), 15–122. The papers sparked lengthy discussions at the meetings in which they were presented. The discussions and subsequent correspondence is included in the *MPICE* pages cited above.

[17] See, by way of illustration of this trend, IOL&R, L/PWD/5/1, Public Works Old Series, Collection 4: 'On the nature and extent of control to be exercised by the Government of India, over the proceedings of the local Governments in regard to Railway matters'; IOL&R, L/PWD/3/69, Bengal RR Letters, no. 81 of 1869, dated 5 Aug. 1869; IOL&R, P/1515, PWD, RR Estabs Progs., June 1880, nos. 3–4.

presidency cities and provincial capitals; and the level involved with the physical construction of a line of railway. The regular civil authorities also had a role and it was they who maintained police and judicial powers over the construction workforce. The relationship between these levels, although hierarchical, was not a simple one of control versus operation, of executive versus line functions. In Britain, for example, substantial controlling functions inhered in the Secretary of State for India through appropriate functionaries in the India Office who, in turn, often acted upon recommendations from the Government of India. Various mechanisms, including the ex-officio Government Director who sat on the board of each private guaranteed company, ensured that this control was extended to those companies. Also, in Britain, consulting engineers to government and private companies advised generally upon railway matters and prepared or vetted the designs and specifications for major works such as great bridges.

The India Office approved the indents for construction material and construction personnel that were translated into substantial orders that went out to British manufacturers and into the recruitment of engineers, platelayers, tunnelers and other experts.[18] This was a major job. By the close of 1863, 2,764,781 tons of railway material—rails, sleepers, locomotives, etc.—valued at £15,128,856 had been sent in 3571 ships from Britain to India.[19] Another source captures the essence of the task of assembly and transportation. Each mile of railway built in India through the 1860s required, on an average, a separate ship carrying some 600 tons of material from Britain.[20] And, having been transported across the ocean, each ton of material had to be moved from port to worksite—a task of enormous difficulty, involving a substantial demand for labour and draft

[18] The substantial benefits to British industry of railway construction in India and the extent to which British labour was a participant in the construction process is captured in the following simple but telling example. *The Engineer*, January–June 1861, p. 64 reported that the Britannia Iron Works had received an order to supply the EIR with 2500 tons of spike to fasten the chairs (in which the rails were secured) to wood sleepers. It was probably the largest, single order of its kind placed till then. Although each spike—a simple item indeed—weighed but one lb., the manufacturing of one spike from pig to completion employed 40 men and boys. Consider, then, the employment and profits generated by a contract to provide the complicated ironwork for an entire bridge.

[19] *The Engineer*, 19 August 1864, p. 120 went on to observe that there was 'no inconsiderable advantage to the parent state' in all of this, with 'great advantage' yet to come as railway construction in India proceeded.

[20] Bánerji, pp. 66–7.

animals in the early years before the growing network of operating railway lines provided some relief.

In India, at the imperial and provincial capitals, the situation was similar to that in Britain in so far as bureaucrats, senior engineers and accountants within the State railway structure or as government engineers supervisory the activities of private companies exercised a close watch over the construction process. This supervision was particularly close in areas of route determination, specifications, costs, progress inspections and approval inspections prior to the opening of a line.[21] In so far as this supervision involved decisions about specifications, about sanctioning or withholding sanction of expenditures, or about the speed and quality of work in progress or completed, it impinged directly on the act of construction. Moreover, it was government's wish to utilize contractors located in India that was important in the EIR's largely unsuccessful attempt to build its line through Bengal and NWP by giving large contracts to Britons domiciled in India, and in the rapidly abandoned attempt to use Indian contractors to build the Madras Railway (hereafter MR). Pressure from the Punjab Government led to the attempt to build the line in Punjab using Punjabi contractors. At a more general level, fiscal decisions in Calcutta and London determined the overall pace of railway construction, which, in turn, largely dictated the number of Indian construction workers employed at any time.

The third level, and for the purposes of this study the most important, was the actual construction of a length of line. Usually a major construction project, i.e. a complete line or a significant addition to an existing line, was placed in charge of a chief engineer (hereafter CE) who was responsible for the construction and the personnel, submitted estimates and designs for the line to the higher authority for sanction, and was responsible for ensuring that the estimates were adhered to or had to justify increases and over-expenditures. The CE usually recommended whether the line should be built by contractors or by the engineers themselves

[21] The Government of Bombay asked for more clerical railway workers in 1863 in the face of pressure from the Government of India to reduce the number of clerks. Bombay bemoaned the amount of labour an audit of a departmentally constructed line involved. 'Masses of elaborate pay sheets have to be examined as to their correctness. All charges for works, stores and establishments again examined as to Government sanction, and again all establishment and pay sheets (the latter fortnightly) are examined for Income Tax, and a role of those subject to it, has to be prepared and forwarded to the Collector.' IOL&R, L/PWD/3/273, Bombay RR Letters, no. 49 of 1863.

acting as contractors, by what was known as the departmental system, or departmentally. As a man of experience, the CE may have been involved in the actual designing of particularly difficult sections and items. The office of the chief engineer was a pivotal one in the entire structure of supervision and management. The CE linked the higher levels of supervision and control (government and also, if a guaranteed company, the agent and board of directors) to the lower levels of project management, whether a substantial contractor was involved or not.

If the construction was being done departmentally, the line was divided further into divisions and sub-divisions headed by junior engineers down the hierarchy. At the bottom were the assistant engineers responsible for 10–40 miles of track. They were physically present at the worksites on a daily basis during the construction season. It was to these assistants that many of the practical manuals on how to build railways were directed. Beneath these engineers were British overseers, who were often former non-commissioned officers or private soldiers of the British Indian Army who had taken their discharge in India, and a number of skilled workmen recruited directly from Britain.

If the line was being built by one major contractor or possibly a number of substantial contractors, the job of the chief engineer and his subordinates—whether the line was private or state-owned—was largely supervisory, to ensure that the contractor did the job properly and in accordance with the specifications set out in the contract. To carry out this supervision, engineers, though fewer in number, were again divided up along so many miles of line. The contractor or contractors, however, had their own personnel who directed the actual construction: British senior personnel, usually styled 'agents', who were often civil engineers and lower-level supervisors, etc. Engineers, not uncommonly, crossed over from the employ of a railway company to that of a contractor and back again, and/or moved into continuing employment in the government PWD.

Two features of the framework that remained constant in substance, though changed in form, throughout the period were: (1) extensive government supervision and control of the construction process; (2) the reliance of both government and private companies at all levels of the process on experts, usually engineers, whose technical knowledge made the building of railways possible. These experts were either employees of the PWD or of private companies, or retained by them on a temporary or an on-going basis. The degree of expertise possessed by the engineers varied considerably. Some assistant engineers at their first construction site had limited experience with railway construction; some of the consulting engineers in England and chief engineers in India were among the most senior and knowledgeable of civil engineers in mid-century Britain, although in the 1850s and early 60s neophytes and long-standing practitioners alike knew little of Indian conditions. An obituary of the CE of the Rajputana State Railway built in the late 1860s and early 70s, one of the first of the State lines, noted that the deceased's staff had been composed of engineers with little or no railway experience. 'His service to the State was thus of great value in training a body of engineers who were the earliest representatives of the important branch of the Indian Public Works Department' of whom some went on to hold the highest posts in that branch.[22]

Thus, the actual construction of a line of railway was supervized and directed through a complex hierarchy of interests and interested parties: politicians and members of boards of directors of private companies in Britain, administrators in India and Britain, supervisory engineers, contractors and their agents, and engineers on the line of works. The hierarchy reached down to the industries and workpeople of Britain who supplied much of the finished or semi-finished materials—and the engineers and skilled workmen—to India. The hierarchy also reached down into India where it ended in a myriad of individual worksites where Indians physically built the railways of the Raj. However, all levels of the construction process and the interaction between and among these levels collectively determined the specific nature of particular construction projects. All construction activity, in turn, represented the application of British industrial capitalism to India mediated through the colonial connection.

Decisions made in London or Calcutta affected the conduct of work at a given construction site from the number of people employed to the tools used or to the degree of danger involved. The consulting engineer in England, Sir Alexander Rendel, who designed the Lansdowne bridge across the Indus at Sukkur, certainly made that job more difficult and hazardous with his awkwardly complicated 3300 tons of steelwork.[23] At the worksites Indians accepted or resisted decisions taken by those who commanded their labour and thus they, in turn, affected the construction process. But what mattered most in terms of getting the railways built was the continuity of the organizational structure that saw individual projects (themselves massive) brought to completion and, decade after decade, the

[22] Obituary, 'Willoughby Charles Furnivall', *MPICE*, 146 (1901), pp. 285–6.

[23] P.S.A. Berridge, *Couplings to the Khyber* (New York: Augustus M. Kelley, 1969), p. 122.

advance of the overall project to develop a railway network in South Asia. Railway building had a management structure that persisted in its essence, regardless of the occupants of particular offices or positions although some individuals undoubtedly did their jobs more effectively than others. Projects were finished, senior civil servant retired, contractors quit or died, chief engineers were fired, assistant engineers were promoted and moved, workers came and went, but, without cessation, although with the occasional slowing of pace, the building of railways continued.

Regardless of subsequent developments and controversies, the decision in late 1849 was to permit private, guaranteed companies to build and operate the railways of India under the close supervision and direction of the Government of India. Specific contracts could be let; lines could be built and opened for traffic. India was about to enter the railway age. The often difficult terrain depicted in map 1 was going to be mastered by a new form of transportation.

The Governor-General from 1848 to 1856, Lord Dalhousie, was committed to railway development. He actively involved himself in deciding where the initial, 'experimental' lines should be built. Based on the success of those experiments, it was Dalhousie, advised by Major J.P. Kennedy, R.E., who established the basic blue-print that was to guide the early decades of railway development in India. Dalhousie's famous 'Railway Minute' of 20 April 1853 carried the day with the Home Authorities on most matters, large and small, including the central issue of the routing of the major trunk lines. Dalhousie also supported the use of private, guaranteed companies under close government control.[24] Earlier decisions by Dalhousie had committed India to the 5′ 6″ gauge and to a land assembly programme for the major lines that would provide for subsequent double-tracking.

Two experimental lines had been approved of in 1849. By mid-1850 Dalhousie had selected the routes and sanctioned the start of the construction of a 121-mile line in Bengal, extending north-westwards from

[24] Dalhousie had served in Sir Robert Peel's administration as President of the Board of Trade, a position that deeply involved him in railway matters in Britain. A good discussion of Dalhousie and railway development in India, including a gloss of the 20 April minute, can be found in M.N. Das, *Studies in the Economic and Social Development of Modern India: 1848–56* (Calcutta: Firma K.L. Mukhopadhyay, 1959), pp. 26–108. The complete text of the 20 April minute, Dalhousie's other minutes and despatches and the railway memoranda of his officials can be found in a number of IOL&R collections or in the Scottish Record Office, Dalhousie Papers, GD/45/6/389 and 411. Davidson, pp. 72–132, goes over similar ground but gives more attention to Kennedy's input.

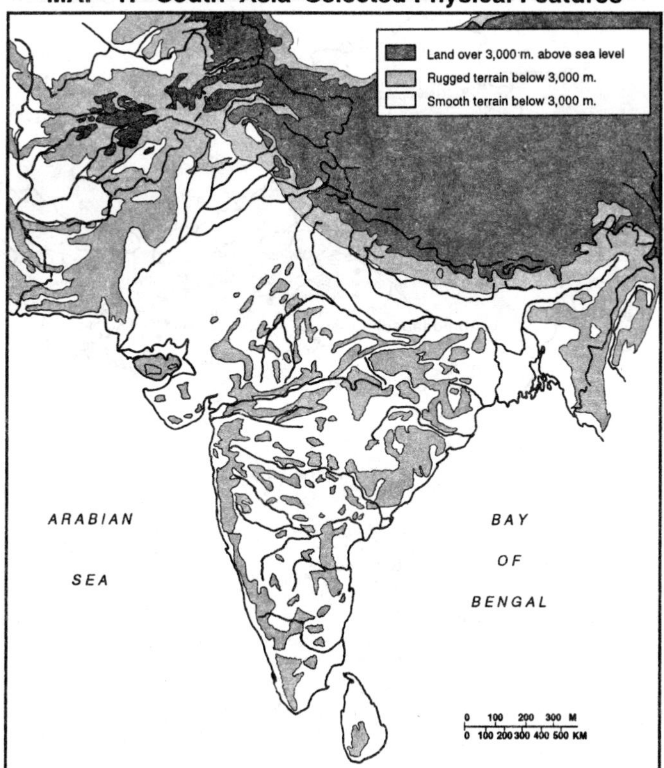

Map 1: South Asia Selected Physical Features

Howrah (across the Hooghly river from Calcutta) to the small town of Raniganj, centrally situated with respect to the coal-fields of Burdwan, and a 35 mile line from Bombay east to Kalyan at the foot of the Western Ghats. Map 2 provides the location of these, and some other, places mentioned in this book. The contracts for these lines—the contracts of August 1849—were made with the EIR and the GIPR respectively. In 1850, survey work was undertaken, tenders were called for the construction of portions of the line and by late 1850 railway construction was underway.

The first in the race to bring the iron horse to India was the GIPR. Construction of the line to Kalyan via Thana—an area now covered by the network of Bombay suburban lines whose rush-hour electric trains, jammed to overflowing with passengers, run back to back at brief inter-

Map 2: Rivers, Sites and Areas

vals—began in the Fall of 1850. The ceremonial turning of the first sod took place on 31 October before a large crowd.[25] The work pushed ahead in 1851. Nine-hour days and work during the months of the monsoon rains took their toll on the contractors' European agents but by mid-1852 some line was open for construction traffic.[26] Directors of the GIPR and their guests were carried to Thana in November 1852; the line to Thana was officially opened on 16 April 1853, when 400 guests travelled on the line with a band and gun salute to start them off at Bombay and a

[25] Das, pp. 62–3.

[26] IOL&R, Photo. Eur. 197: G.O. Mann Collection. Mann to his Father, 30 October 1851; J.N. Westwood, *Railways of India* (Newton Abbot: David & Charles, 1974), p.22.

refreshment tent to greet them 75 minutes later at Thana.[27] By May 1854 the entire line to Kalyan was in operation. Each official opening was marked by an appropriate ceremony with various festivities, banquets, speeches and toasts, plus the necessary train ride in the foreshadowing of a ritual that was to be repeated throughout the century in various parts of India as more and more lines were opened, great bridges completed, long tunnels excavated or precipitous inclines surmounted.[28]

Meanwhile, some 1000 miles to the east of Bombay, near Calcutta, the EIR similarly proceeded with its experimental line. In September 1854, 37 miles were officially opened and the entire 121 miles to Raniganj were formally opened on 3 February 1855.[29] Sadly, Dalhousie, the architect and director of the first phase of Indian railway development, incapable of the journey to Raniganj, had to content himself with an opening day visit to Howrah. One of the most hard-working of the Governor-Generals, he was, by 1855, worn-out and very ill. He returned to Britain in 1856 and, never regaining his health, died well before his planned railways fully materialized.[30] He was not the first victim, direct or indirect, of Indian railway construction, nor was he to be the last, but he was one of the founders of the network that was to eventually spread throughout the subcontinent. In the meantime, the experimental lines he had sanctioned in 1850 seemed to be well on their way to meeting the objectives he had set for them: 'to prove, not only that it is practicable to construct railways in India, as engineering works, but that such railways, when constructed, will, as a commercial undertaking, offer a fair remunerative return on the money which has been expended on their construction.'[31] Dalhousie can be excused for not knowing that it would take some fifty years for most lines to become remunerative but he did know lines could be built and operated. The blueprint of 1853 was based on what was

[27] J.N. Sahni, *Indian Railways One Hundred Years, 1853 to 1953* (New Delhi: Government of India, Ministry of Railways, Railway Board, 1953), p. 3. On 17 April the famous, wealthy Parsi merchant and philanthropist Sir Jamsetji Jijibhai rented the entire train to convey him and his family to Thana and back.

[28] M. Satow and R. Desmond, *Railways of the Raj* (New York and London: New York University Press, 1980), p. 13 provides excerpts from the newspaper coverage of these early GIPR openings.

[29] Davidson, p. 153. A description of the February 3 ceremony can be found in [G. Huddleston], 'The Opening of the East Indian Railway', *Bengal Past and Present*, 11:1 (January–July, 1908), pp. 55–61.

[30] The standard biography is Sir William Lee-Warner, *The Life of the Marquis of Dalhousie, K.T.*, 2 vols. (London: Macmillan and Co., 1904).

[31] Dalhousie's minute of 4 July 1850 as quoted in Sahni, p. 9.

known and on what the optimistic proponents of railways for India felt was sure to happen—commercial success.[32] One by one the lines proposed by Dalhousie received official sanction and their construction began.

The central elements of Dalhousie's planned railway map of India were the trunk lines connecting the major administrative centers of the presidencies and the provinces. The line from Calcutta would strike north-westward up the Ganges valley to Allahabad, and then on to Agra, Delhi and beyond into the newly-conquered Punjab and its capital, Lahore. A line from Bombay would strike north-eastward, by a route to be subsequently determined, to join up with the line in the Ganges valley, thus providing a Bombay to Calcutta connection. Railways were also to be built from Bombay into Khandesh and to the city of Poona. Lines from Madras City were to be built to the western coast of the Madras Presidency with branches to Bangalore and the foot of the Nilgiris, while another line was to strike north-westward through Cuddapah and Bellary with a view to a further extension to Poona, thus creating a Madras to Bombay connection.[33]

Dalhousie died, aged 49, on 19 December 1860. If, in his final days, he reflected upon Indian railway development and his role in stimulating and guiding that effort he must have derived satisfaction from what he saw. Hundreds of miles of line were open for traffic and nearly 3000 miles were under construction or about to be commenced. Railways were being built in many parts of the subcontinent; lines beyond the plans of Dalhousie were being started. Map 2 indicates the lines in operation in 1861, plus those subsequently opened and in operation in 1881 and 1901.

Construction in the Madras Presidency began extensively in early 1856, and by July of that year, the easy 64 miles from Madras to Arcot, started in 1853, had been completed and the line opened for traffic.[34] Subsequent development was slower, but by late 1860, the MR had some 136 miles of operating line and 548 planned or under construction, while the more recently sanctioned Great Southern Railway of India (hereafter GSIR and later SIR, the South Indian Railway) was nearing completion of its 78-mile connection of Negapatam and Trichinopoly.

[32] Though in the last half of 1853 the Bombay-Thana line returned 4 per cent. Das, p. 85.

[33] Das, pp. 75–6; Davidson, pp. 88–94.

[34] An official sod-turning ceremony had taken place in June 1853, but work then was placed on hold while the overall direction of railway development was determined. Davidson, pp. 345–6.

Meanwhile, the work of the first companies had pushed ahead. The GIPR had opened 297 miles and was pushing ahead with the construction of a further 787 miles up and beyond the immensely difficult Western Ghats. The EIR had 368 operating miles of track, and an additional 761 in course of execution or sanctioned.[35]

Further work was afoot in Sind and Punjab where the Sind, Punjab, and Delhi Railway (hereafter SP&DR) was building one line north from Karachi and lines south-westward and eastward from Lahore towards Multan and Delhi respectively for a total of 364 miles. The Bombay, Baroda and Central India Railway (hereafter BB&CIR) had started, in 1855, to build a line north from Surat into Gujarat and to the city of Ahmadabad. Once the hopes for revitalizing Surat as a major harbour—as it had been in the seventeenth century—proved unrealistic, the BB&CIR obtained permission in 1859 to build southward from Surat along the coast to Bombay. Thus, by the end of 1860, this company had 35 miles in operation and 219 miles of line under construction.

The list of active railway companies at the close of 1860 is completed by the addition of two railways under construction in Bengal. The Eastern Bengal Railway (hereafter EBR), 110 miles from Calcutta to Kushtia, represented the start of a project intended eventually to carry on to Dacca. The 29-mile Calcutta & South-eastern Railway was designed to run from the imperial capital to Canning, a new port on the Matla river, in the hope that it would attract the cargoes of Calcutta-bound ships wishing to avoid the dangerous navigation of the Hooghly river.

The first decade of railway building in India exhibited most of the problems and obstacles that were to bedevil the construction process during the subsequent four decades. Foremost among these were: the human problems associated with obtaining, retaining and effectively utilizing large bodies of labourers; and, the technical problems associated, in particular, with the task of building bridges over rivers whose water-flows in wide, sandy beds changed from benign dry-season trickles to vast rainy-season torrents that could quickly scour beneath all but the deepest of bridge foundations. But in reality these were sides of the same coin. The social organization of labour and the work itself took place through means of production defined by the existing technologies of India interacting with the transplanted technologies and techniques of British contractors,

[35] Information on the miles of line open and under construction or sanction at the end of 1860 is compiled from *Railway Report 1860–1*, p. 6, and *Railway Report 1869–70*, p. 5. Routes can be followed in Davidson, *passim*.

their agents and the companies' engineers, which, again, were constrained and contained by what Indian labourers could and would do, what they knew how to do and what they could be trained to do.

Thus, Henry Fowler, who, with his partner Faviell, had the contract to build the first portion of the GIPR line, wrote to England in May 1851 that he was trying to get workmen to start at 6.00 a.m. rather than the customary 8.00 a.m., but 'it is a most difficult thing to alter existing systems as almost every custom the natives have is founded on absurd but invincible prejudices—generally of a religious character'.[36] Fowler went on to note that men were divided into castes and would only do a particular kind of work and often would not work alongside men of another caste. Finally, he recounts how he was one day on the line of works at 10:00 a.m. when it was already very hot, so he took a workman's water-pot and poured the contents over his head only to see the 'innocent vessel . . . immediately doomed to destruction as the fact of my touching it had defiled it'.[37] This little story illustrates a British perception of the problems associated with the organization and utilization of Indian labour, and it reminds us of the differences that existed at the construction interface: cultural differences between workers and their European bosses, and differences between various groups of workers. However, real as these differences were, one should not over-emphasize their importance in the labour process. The first British railwaymen in the field found Indian work practices peculiar and obstructive. Later railwaymen both adapted to those practices and modified them. The adaptations and modifications have been explored in a later chapter.

Other problems that re-occurred throughout the period from 1860 to 1900 also appeared in the decade 1850 to 1860. The relationships between government officials and those building the railways—relationships that existed at a multiplicity of levels from the comfortable offices of the highest authorities to on-site encounters in the heat and aggravation of the lines of work—were prickly. The health of workmen, European and Indian alike, was a constant worry as high levels of morbidity and mortality ravaged the workforces. Cholera epidemics were the great, known scourge; malaria, its cause and prevention still unknown, the more insidious and slower acting danger. The timely delivery of supplies was

[36] IOL&R, Eur. Mss. C. 401, Fowler to Leather, 2 May 1851. The attempt to lengthen the working day is a standard managerial manoeuvre to increase labour output under the conditions of formal subsumption.

[37] Ibid.

another problem. Much of the construction equipment and material came from Britain and its date of arrival at the work place was often tardy.[38] One item that was frequently obtained in India, wood for the sleepers (ties), was difficult to obtain in sufficient quantities and at the needed time. Moreover, a process of experimental trial and error went on to determine just what kind of plant, material and techniques were best suited to Indian conditions. Engineers had to grapple with problems as diverse as bridge design and the selection and treatment of wood best suited for sleepers, given India's climatic conditions and the ubiquitous white ants (termites).

These problems continued to be present in subsequent decades, but they gradually became less and less bothersome; the problems and obstacles were increasingly well known and expected, the solutions increasingly efficacious. Indeed, at one level, the period from 1850 to 1900 can be seen as one in which railway construction in India progressed from the point where much was novel and demanded ad hoc solutions, to the point where the construction process in many of its aspects had become routinized and based on substantial proven experience. Supplies of labour became more assured and skilled bodies of workmen came into existence, who moved, particularly where bridge work was concerned, from one project to another. Bridge foundations became more secure and bridges stopped collapsing; the causes of epidemic mortality and its diminution became better known, although malaria was not tackled effectively before the twentieth century; railway builders learned to live with government supervision and the government, in any case, started to build and operate railways on its own account.

The routinization of railway building in India can be documented, but also suggestive is the increasing absence of revealing, detailed construction information of the kind one finds in the earlier government records. The relevant Proceedings volumes of the Government of India and its presidencies and provinces (usually in the Public Works Departments) throughout the 1850s, 1860s and 1870s are full of information relating to all aspects of railway construction, but as one reads the volumes of the 1880s and 1890s, the information becomes sparse, and what there is is less descriptive, less phrased in terms connoting novelty and challenge, and more and more standardized and quantified.

[38] About one per cent of the 5703 ships that sailed to India with railway goods up to 1868 were shipwrecked. Banerji, p. 66. Insuring these shipments was another indirect benefit of Indian railway construction to British companies.

Routinization came from incremental advance. There was no dramatic breakthrough; just the application of accumulating experience preserved in memory and print, transmitted in part through an increasing corpus of engineering manuals, papers in professional journals, series of technical papers and through instruction in educational institutions. Indeed, entry into the railway age stimulated the rapid development of a professional engineering literature in India. Publication of *The Engineer's Journal and Railway and Public Works Chronicle of India and the Colonies* began in Calcutta in January 1858, while *The Bombay Builder. An Illustrated Journal of Engineering, Architecture, Science and Art* first appeared in July 1865.[39] The staff at Thomason College, Roorkee, published the various series of the *Professional Papers on Indian Engineering* between 1863 and 1886. The Office of the Director of Railway Construction (within India, PWD Railways) began its series of Technical Papers around 1890; this series, continued by the Railway Board after its creation in 1905, is still published by the Board in India as the twentieth century draws to an end.[40] Thomason College, Elphinstone College, and other institutions joined later by places like the Sibpur Engineering College, trained many of the overseers and, later, some of the first—and for long the very few—Indians who became professional engineers. In Britain, engineers were trained at a variety of institutions, including from 1871 to 1906, the Royal Engineering College at Cooper's Hill from which so many of the British-recruited engineers for India graduated. The British professional engineering literature also provided an important outlet for papers reflecting engineering challenges and experiences in India.

Presumably, a corpus of preserved experience was also acquired by skilled Indian workers, although by a process that, by its nature, left no records. Indians learned the skills of railway construction by precept, initially in some cases from skilled British workmen and engineers, and by experience. The skills were then transmitted orally among the Indian gangs of skilled workers and from them to their children. Such would have been the expertise 'acquired from long experience' by Indian miners driving through the exceedingly hard quartz rock encountered in the excavation of the Monghyr tunnel or by the workmen, or their descendants, who moved for decades in North India from one great bridge-

[39] The BM has runs of both covering, respectively, 1858–69 and 1865–9.

[40] The series contained 297 items by the time a catalogue was published in 1925. See IOL&R V/27/2/14, Govt. of India, Railway Department, *Catalogue of Technical Papers* issued by Technical Section of Railway Board of India, 1925 (Delhi: Government of India Press, 1925).

building project to the next.[41] The 1850s and 1860s witnessed the recruitment and training of the crucial first generation of skilled railway builders.

The process of routinization also demarcates the 1850s—and probably much of the 1860s as well—from the later decades. The people who worked in the 1850s and the early '60s to build the first railway lines, were the pioneers—some succeeded, some failed and many died in the attempt. The pioneers began the construction, and thus they began the process of creating the experiential foundation upon which the subsequent, more settled advance was based. Initially, few precedents existed for constructing railways in social and physical conditions like those present in India.

In one respect, however, the later 1850s were singularly different for railway builders, at least for those in much of northern India. The Santhal rebellion of 1855–6 followed by the mutinies and civil uprisings of 1857–8 affected railway construction to a greater or lesser extent in the stretch from Rajmahal to Delhi—the Santhals provided the earlier disruptions in the East; the events of 1857–8 had a greater effect on works underway in the mid-and upper-Ganges valleys. For the main part it was the EIR that was affected. Work was generally halted, work gangs dispersed, station works and some bridge works destroyed, Europeans in the field forced to flee for safety or to stand and defend themselves.

The Santhal uprising of 1856, touched of in small part (the Santhals had many and longer-standing reasons to be aggrieved) by the oppression of European railway builders in the Rajmahal hills, was brutally suppressed by British-led troops.[42] Heroism with bows and arrows and axes was no match for the firearms of the sepoys, as the death toll of some 20,000 Santhals attested. As for the 'Mutiny', we can appreciate the unintended irony in the following statement by the Viceroy, Lord Canning, in 1860:

[41] Railway Letter (no. 12), 18 December 1860, Governor-General [Canning] to Sir Charles Wood, Secretary of State for India, reproduced in *Railway Report, 1860–1*, p. 29.

[42] Kalinkar Datta, *The Santal Insurrection of 1855–7* (Calcutta: University of Calcutta, 1940), esp. p. 8. A.F.C. De Cosson, 'The Early Days of the East Indian Railway. A Side-Light on the Mutiny', *Bengal Past and Present*, V (January–June 1910), p. 264 asserts that the charges against the contractors (Nelson and Co.) proved to be without foundation. Ranajit Guha, *Elementary Aspects of Peasant Insurgency in Colonial India* (Delhi: OUP, 1983), p. 143 cites official documents that charged the railway builders with bullying Santhal labourers and disgracing Santhal women. Nelson and Co. certainly proved willing to help the authorities suppress the Santhals (see Datta, pp. 87–92).

At the Soane Bridge, and near to it, I observed amongst the workmen many who had evidently been Sepoys. Many of these had openly received their discharges from the Army, but there were others who had probably slunk back to their homes unobserved. They were spoken of by the Engineers as useful, well-behaved men, and as being those from whom the police of the works and petty officers were mostly chosen. This employment of a class who are not generally willing to take to field labour in their own villages has, I have no doubt, contributed to the thoroughly peaceful condition which now prevails throughout Bihar.[43]

The main effects of the various uprisings and disturbances were to slow the progress of construction of the EIR and to increase its cost. Company officials tried to estimate the delay caused by the events of 1857–8 and concluded that it ranged from a setback of 35 months in the area close to Delhi, 30 months on the Soane Bridge works, twelve months in and around Cawnpore, nine months in several localities, and so on, down to six months in the Patna area.[44] The magnitude of the cost increases was harder to estimate since they involved both direct and indirect factors. The destruction of construction plants and of completed or partially completed works (station buildings, bungalows, bridges and their approaches) represented additional direct costs. Delay attributable to the disturbances represented additional indirect costs, some of which were lost opportunity costs. European engineers and overseers had to be paid. Carters, impressed into the service of the avenging armies of the Raj, had to be re-hired and induced, sometimes at higher rates, to return to the construction areas. 'Some thousands of carts', the engineers complained, 'were taken from the neighborhood [around Hallohar] and all the way up to Patna for the use of the troops'.[45] Labour, it was argued, was more expensive after the 'Mutiny'. The daily rates for ordinary carpenters rose from a range of 3–4.5 pence to 4–5 pence; masons went from 2–2.5 pence to 5 pence; divers from 6 pence to 8 pence.[46] Material such as timber, lime and firewood also increased in cost. These increases were attributed to the disturbances, but some of them might have occurred in any case since railway construction, especially in the first decade, sometimes created a demand for labour and materials beyond the immediately available supply. Regardless of that possibility, the authorities estimated the cost

[43] *Railway Report, 1860–1,* Canning to Wood, 18 December 1860.

[44] IOL&R. L/PWD/2/75: Railway Home Correspondence—C. Register I—East Indian Railway Company, 1859. Enclosure to 143/59.

[45] BM. IS 180/2. East Indian Railway, series of half-yearly reports, 1858–60: 'Report of George Turnbull, Chief Engineer', 18 August 1858, p. 8.

[46] IOL&R, L/PWD/2/75/, EIR 1859, enclosure to 143/59.

increases attributable to the 'Mutiny' to be 20–50 per cent, varying with locality.[47]

The events of 1857–8 continued to have an impact on railway construction into the 1860s. At one level the Mutiny reinforced the determination of government to push ahead with trunk line construction. Operating railways, it was recognized, were a powerful tool for ensuring the security of the Raj; they greatly facilitated the movement of troops. Thus, if the costs of railway building were a major concern of government in the 1860s, this was balanced by an appreciation of how important they could be for the maintenance of British rule.[48] Related to this issue of security was the concern, at times an obsession, that was to last for decades among the authorities, namely ensuring the military security of the railway lines, bridges, tunnels and stations.[49] The most tangible expressions of this concern were the decisions to design various stations as fortified, defensible buildings; to fortify the approaches to major bridges; and to adorn tunnel entrances with crenelated towers.[50] Would one be too imaginative to see a similar purpose in the outwardly grim, brick, fortified, southern Gothic cathedral of St. Cécile at Albi in France and the grim, brick, fortified, main railway station at Lahore?[51] The guardians of the established order, shaken by 'heresy' and by 'rebellion', emerged triumphant, yet still fearful of those they had sanguinarily suppressed; they built these two great edifices as massive assertions of an order uneasily restored. The symbolism was the same despite the leap from the fourteenth to the nineteenth century, and from Albigensian 'heretics' to Indian 'rebels'. At quite a different level, the engineering teams were still receiving military escorts in the Jubbulpore area in the Spring of 1860 and contractors' claims for losses they had suffered during the uprisings were still being processed.[52] Engineers were receiving medals and other recognition for

[47] Ibid.

[48] See, for example, IOL&R, P/217/37, NWP, PWD progs., Railway Branch, April 1864, no. 83 for a statement of the military and political importance of the Delhi–Amritsar line; also the need for defence of the bridge at Delhi.

[49] See, for example, IOL&R, P/191/15, PWD Progs., Railway, June 1865, nos. 57–8 on the defence of railway stations.

[50] See the photographs in Satow and Desmond, pp. 57 and 86 of the mouth of the Khojak tunnel and the Lahore station respectively.

[51] Both structures, too, were major representatives of the central icons of their respective ages.

[52] IOL&R, NWP, PWD progs., Railway Branch, April 1860, nos. 66–79; May 1860, nos. 10–12; June 1860, nos. 103–5.

their service during the 'Mutiny'; others were arguing that their service had been equally meritorious and they too should receive medals.

With the events of 1856 to 1858 behind them, the builders of India's railways returned to their efforts in the north and more actively pursued their works in the south, west and east. Some 3000 miles of line were being planned or actively constructed in 1859. Extensive construction continued for the remainder of the century. Table 1 displays, in one-year intervals, the total route miles open for traffic from 1853 to 1900, the additional number of miles opened each year between 1853 and 1900 (the result of subtracting from each year's route miles open the preceding year's route miles open), and an estimate of the number of miles under construction in each year between 1859 and 1900. The impact of the disturbances discussed above makes it unwise to estimate miles under construction prior to 1859. The column headed 'total route miles open' enumerates the progress of construction; it displays the results of the construction process. The route miles of track in operation grew from 20 in 1853 to a substantial 8995 in 1880, and to an impressive 23,627 in 1900. Map 3 and the graph below, provide a visual depiction of the growth. Had the table been continued into the twentieth century it would have displayed an additional 6945 miles added in the first decade, leading to a total mileage, in 1910, of 30,572, when India's railway system was the fourth largest in the world.[53] Some route miles were double-tracked—800 miles by 1880 and 1474 by 1900—that could be added to the route miles totals to give total running miles. Many of the first trunk lines were built to accommodate double-tracking, although the second track was not necessarily installed at the initial construction. These lines required more work since embankments, tunnels and cuttings had to be wider, bridges and culverts stronger.

Construction proceeded steadily throughout the half-century. On an average, 502 route miles were added annually to the operating network. A percentage growth rate, however, better captures the nuances of the progress. In the period 1861–1900 (using 1861 to avoid the distorting effect of the 749 miles opened between 1860 and 1861), the annual percentage growth rate was a solid 7.2 per cent—a figure that better catches the overall performance for the bulk of the period (39 of 47 years) than does the 16.2 per cent for the period 1853–1900, which is influenced by the low values but high growth rates in the early years between 1853 and 1860, when the miles open grew by an average annual rate

[53] Hurd, 'Railways', in *CEHI*, p. 737. Route miles totalled 40,524 in 1946–7.

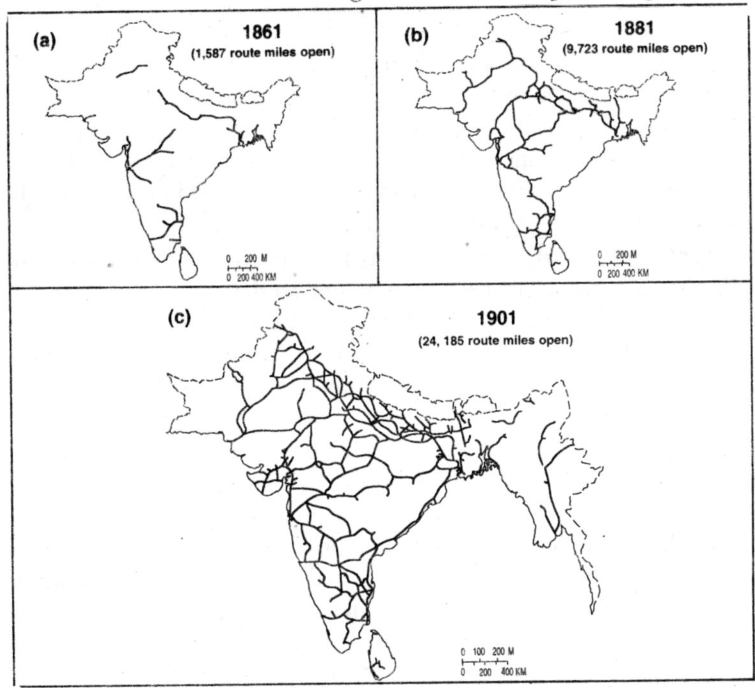

Map 3: Growth of the Railways, 1861–1901

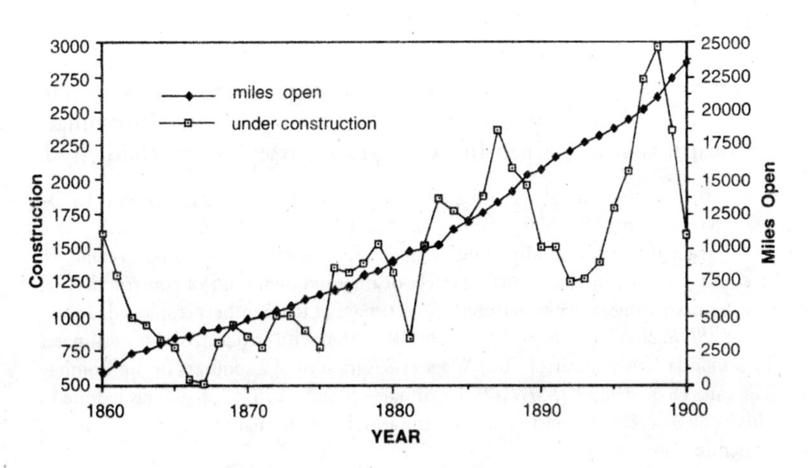

Graph: Miles open and under construction, 1869–1900

of 70.5.[54] Moreover, although the percentage rates are useful for comparative purposes, it was the absolute number of miles under construction that largely determined the level of demand for construction labour. The graph displays the fluctuations in the growth of the railway network better than these numerical statements of central tendencies. Readers are reminded to be aware of the two vertical axes, while interpretating the graph.

Unfortunately, the railway records do not provide an annual statement of miles actually under construction, nor do the more detailed records provide the disaggregated data from which a pan-Indian time series of miles under construction could be brought together. The table summarizing construction progress in each annual *Railway Report* usually had a column headed 'length remaining to be finished'.[55] Unfortunately, this column normally included all mileage sanctioned (i.e. officially approved) for construction, which meant that the miles in question could have been at any stage of the construction process—from a paper existence whereby the miles in question were approved but the project would not start for a number of years, to completed but not officially opened. As forward planning lengthened in the later decades, 'length remaining to be finished' increasingly represented miles planned and sanctioned but not yet under active construction. As one *Railway Report* stated: 'A mileage rate of cost for each province cannot be correctly assigned from these figures, because many of the lines of railway are in various stages of completion.'[56]

However, if the figures do not exist, and if it is desirable to have annual, pan-Indian estimates of railway construction employment—and for some of the arguments advanced in this book such estimates are not only desirable but necessary—then some surrogate measure has to be used to approximate the miles under construction annually. Without an estimate of construction miles, a pan-Indian employment series cannot be derived.

[54] Other average annual per cent growth-rates were: 1860–70 = 19 per cent; 1870–80 = 6.6 per cent; 1880–90 = 5.9 per cent and 1890–1900 = 4.1 per cent.

[55] I used this information, transformed into a three year moving average, to account partially for the distorting effect of the unknown stage of construction, in my 'Constructing railways in India—an estimate of the numbers employed, 1850–80', *IESHR*, 20:3 (July–Sept. 1983), pp. 317–39. Useful criticism of those estimates in Sunanda Krishnamurty, 'Real Wages of Agricultural Labourers in the Bombay Deccan, 1874–1922', *IESHR*, 24:1 (January–March, 1987), 81–98 has led me in this book to a reconsideration of my estimates. I present that reconsideration in the appendix.

[56] *Railway Report, 1880–1*, p. 5.

The 'total route miles open' column in Table 1 suggests a way forward. This is a reasonably accurate body of information; it states what had happened rather than what might have been happening. We know, for example, that 749 miles were opened between 1860 (when 838 miles were in operation) and 1861 (when 1587 miles were in operation). These 749 miles had to be at some stage of construction during 1860. In addition, most miles of railway line were constructed over more than one working season. Two working seasons were a standard minimum. The embankments or cuttings were constructed during the first season and, after an earth-consolidating rainy season, repairs and platelaying took place during the second working season. Work on bridges, tunnels and large-scale cuttings or embankments—the most difficult stretches with the largest labour inputs—continued for 3–5 years. Furthermore, work in the 1850s and the early 1860s (some very difficult) appears, on the whole, to have taken longer to complete than similar distances in the 1880s and 1890s. Moreover, a substantial amount of heavy railway reconstruction went on from the mid-1860s as the earlier work proved defective and sometimes collapsed. Floods and other natural disasters also took their toll. This form of railway construction, for which virtually no employment figures are available, does not appear in Table 1 and its absence almost guarantees that the employment estimates presented in Table 2 are not overestimates. Reconstruction was a significant employer of construction labour.

I suggest, therefore, that an estimate of the miles under construction for any given year can be generated by assuming 2.5 years of construction per mile. Thus, to continue the example, we can assume that the additional 749 miles opened between 1860 and 1861 were under construction in 1860 and in 1859. Likewise, the 746 opened between the 1861 count and the 1862 count were under construction in 1861 and 1860. The half-year (the 0.5) is best attributed to the year for which we wish to estimate the construction mileage in order to recognize the fact that the final stages of railway construction usually used less labour and to provide some adjustment for the idiosyncrasies of the *Railway Reports* changing reporting dates (31 December and 1 March were favoured at different times). Thus, the estimate for any given year comes to be composed of the sum of three numbers: the miles opened during each of the following two years plus 50 per cent of the miles opened in the current year. For example, the miles under construction in 1860 were 1602, i.e. 107 + 749 + 746 = 1602 (50% of the 1859–60 opening [213/2 = 107] plus the openings during 1860–1 [749] and 1861–2 [746]. The results of this process are summarized in the 'miles under construction' column in Table 1. This process results in conservative estimates and if there is a consistent error it is, at least in the early years, to underestimate the number of miles under construction.

But how many people were used to build the miles of railways represented in Table 1? The answer to this important question requires an estimate of the average (mean) number of workers employed to construct one mile of railway in nineteenth-century India. Such an estimate, joined to the estimates of miles under construction in Table 1, provides the estimates of the annual size of the railway construction work-force provided in Table 2.

In so far as Indian employment is concerned, I estimate that, on an average, between 126 and 155 workers were employed per construction mile. It must be emphasized, however, that the 126 to 155 range is an aggregate and summary estimate around which there was much variation. Facile application of my summary estimates to particular stretches of construction should not be undertaken without a good understanding of local conditions. I derived the much smaller estimates of the size of the European (largely British) employment in a different fashion. The appendix provides an extended discussion of how the estimates for both groups were developed and a presentation of the data upon which the estimates were based. This book can be read with limited reference to the appendix—absolutely crucial data are incorporated into the text or your attention is drawn to particular tables in the appendix—but the data presented there may be of interest for purposes other than my concern with railway construction. The data from specific sites certainly illustrates the extent to which variations existed around my proposed summary estimates of employment per construction mile. In appendix table 9, for example, one finds daily averages per construction mile of the EIR in the North Western Provinces in 1860, ranging from 4 people to 2067 or, removing the extreme outliers, from 27 people to 443. The richness and variation in the site data are its attractions—it provides a local and regional perspective within this all-India study.

Tables 1 and 2 and the graph display certain interesting features of railway construction in nineteenth-century India, most notably fluctuations in the construction effort. The intense activity of the late 1850s and early 60s, probably understated in Table 1, was followed by less activity during the remainder of the 1860s. Activity intensified in the mid-1870s and, after a brief fall-back, resumed at new and generally higher levels from 1882 onwards. Indeed, the two heaviest construction years came almost at the century's end: 1897 with 2732 miles and 1898 with 2962 miles. Table 1 also indicates that there were sharp annual fluctuations in the number of miles under construction. The 'miles under construction' column has a mean value of 1405 but the standard deviation is a substantial 576 and the range extends from 509 miles (1867) to 2962 miles (1898).

The two most intense periods of construction are readily explained. The late 1850s and early 60s represented the forceful implementation of Dalhousie's grand scheme for an all-India trunk line network that had been delayed or set back by the events of 1857 and, in any case, required a number of start-up years before construction could hit high gear. The activity after 1881 followed a shift in government policy during Ripon's Viceroyalty. Sparked in part by the recommendations of the Famine Commission of 1880 that India needed 20,000 more miles of railway, 5000 of which were needed immediately to protect Indians from famine, the government pursued a more vigorous policy of railway expansion. This policy once again favoured the use of private companies rather than construction by the State, popular in the 1870s.[57]

The other periods of substantial construction activity or contraction are also explained largely by the vicissitudes of government policy—in turn often determined by the financial health of the Government of India—and to the fact that major new lines were sanctioned as complete systems. Thus, for example, the decision in 1871 to build the 500-mile Indus Valley State Railway (hereafter IVSR) soon added many miles to those under construction.

Other causes of the considerable fluctuations in the yearly totals were the fact that different lines or sections of line were built with greater or lesser rapidity—depending on terrain and on the decision of those in charge to push ahead or to go easy—and on the continual sanctioning and completion of branch lines. Regardless of the fluctuations in construction activity, Table 1 demonstrates the sustained and, in most years, considerable effort that went into railway construction in nineteenth-century India. But who managed the work of construction? This chapter has described something of the general patterns of ownership and supervision that characterized the Indian situation. The next chapter looks at the group who was more directly involved in the co-ordination and direction of the construction process—the engineers and contractors.

[57] Sanyal, pp. 135–57, 192–3.

CHAPTER 3

Contractors, Engineers and Petty Contractors: The Varieties and Problems of Management

Lured by the guarantee, British investors made millions of pounds available for railway construction in India. Within two decades of the start of private railway construction, the Government of India began to make millions more available via direct State investment in the construction and ownership of railways. However, capital investment, be it from private sources or from the government, did nothing by itself. The millions of pounds made available for railway construction had to buy labour power. People had to be mobilized and organized to do everything from the basic task of moving earth to the complex task of designing bridges that would span great rivers and support the weight of fully-loaded freight trains. The means of production—tools, machinery and materials—also had to be assembled.[1] Who did this?

The answer, in Britain, in the 1850s was: the contractor. An individual, or more typically a partnership of two or three individuals, who took a contract to build a particular line according to certain specifications, at a certain price. Jack Simmons, a historian of the British railways, refers to contractors as 'a group that stands between the capitalists who found the money for them, the engineers who planned and directed them, and the navvies who, in physical terms, built them—stands between them, and yet partakes of the character of them all. In the building of a railway, no men were more important than the contractors; the whole execution rested on their skill, their organization, and their judgement'.[2] These large-scale contractors had subordinated or replaced the many small

[1] The land needed for the right-of-way was assembled by Government.

[2] Jack Simmons, 'Introduction' to *Life and Labours of Mr. Brassey* by Sir Arthur Helps (reprint edn, New York: Augustus M. Kelley, 1969), p. xvi. Also see the useful chapter on British contractors in David Brooke, *The Railway Navvy* (Newton Abbot: David & Charles, 1983), pp. 73–107.

contractors who had built Britain's canals and earliest railways.[3] Large contractors, like Thomas Brassey, Sir Samuel Morton Peto or Joseph Firbank, came to possess substantial stocks of construction 'plant' (equipment, i.e. fixed capital), an experienced staff of agents, engineers and overseers, and a following of navvies who were prepared to tramp from one project to another. The need to keep this stock of physical and human resources employed led the great contractors to search unceasingly and world-wide for new contracts. In the 1840s and 50s, Brassey and his associates built railways in France, Italy, Austria, Norway, Holland, Spain and India.[4]

Thus, the model of railway building most familiar to the directors and engineers of the guaranteed companies that had won the rights to build and operate India's first railways, involved the use of contractors who would undertake to complete substantial sections of line. Berkley of the GIPR put it most forcibly:

> I found that all the various duties connected with public works [in India] were thrown upon individual officers, so that by making them the executors as well as the designers and superintendents of the work, they deprived the country of the services of a class of honourable men—a class which is foremost in enterprise and in practical experience—and without whose co-operation I believe that many of our noblest works would never have been carried out—I mean the Contractors of England. I at once determined, that as far as it rested in me, my share of the railways of Western India should be constructed by the agency of contractors. . . .[5]

Not surprisingly, therefore, it was to substantial contractors that the first two companies into the field, the EIR and the GIPR, turned. The EIR directors had to turn down a bid by a Mr. Jackson of London in August

[3] B. R. Mitchell, 'The Coming of the Railway and United Kingdom Economic Growth', *Journal of Economic History*, 24:3 (September 1964), p. 323.

[4] Sir Arthur Helps, *Life and Labours of Mr. Brassey 1805–1870* (London: Bell and Daldy, 1872), pp. 154, 161–4; Harold Pollins, 'Railway Contractors and the Finance of Railway Development in Britain', in M.C. Reed (ed.), *Railways in the Victorian Economy. Studies in Finance and Economic Growth* (Newton Abbot: David & Charles, 1969), pp. 217–18.

[5] From a speech given by Berkley at a testimonial dinner in his honour in April 1856 while he was on leave in England as quoted in *The Engineer* (25 April 1856), p. 233. The first part of the quotation represents Berkley's attack on the departmental system which had hitherto largely prevailed in PWD projects. It appears that the adoption of the contract system on many railway projects led to wider use of the system in works other than those of the railway companies. See *The Engineer* (12 February 1869), p. 121.

1849, to construct the first 70 miles out of Calcutta at £8000–9000 a mile when the East India Company Court of Directors and the Board of Control rejected it probably because they thought the tender to be too high—a mistake of giant fiscal proportions given the subsequent costs of railway building in India.[6] Thus, with an unfortunate example of what the government supervision and control clauses in the railway company's contract might entail, the EIR decided to seek men primarily in India who would take contracts to build sections of the experimental line to Raniganj. Among the contractors, whose names are often the only thing we know about them, only three succeeded: Burn & Co. and Norris & Co. of Calcutta, and Hunt, Bray & Elmsley of London.[7]

The failures did not surprise the consulting engineers to government, who knew some of the men to be undercapitalized and whose contracts stipulated too short a period for completion.[8] One person, a Mr. Ryan, was reported to misunderstand completely his position as a contractor.[9] Obviously these were not the large-scale contractors of the British model; a fact tacitly acknowledged in the contracts themselves which reserved to the EIR and its engineers the more difficult task of constructing bridges.

Failures, or the recognition that the Europeans coming forward were not railway contractors of the mid-century model, but counterparts of small, inexperienced contractors of early British railway construction, did not deter the EIR from subsequently tendering extensive sections of line. As the line was sanctioned beyond Raniganj—through the Rajmahal hills, up the Gangetic valley to Patna and Allahabad, then into the Doab to Agra and Delhi—the initial attempt at construction was by sizeable contract, with bridgework again usually reserved for the EIR's engineers.[10] Within the boundaries of the Bengal presidency (the Bengal division of the EIR),

[6] Edward Davidson, *The Railways of India* (London: E. & F.N. Spon, 1868), p. 146.

[7] Pollins, p. 212, writes: 'The contractor is an elusive figure in the history of British railways.' In the history of Indian railways many of the contractors are so elusive as to be historically invisible or, at best, known only as a name. Burn and Co. deserves further study. The firm was in existence in the early 1830s in Calcutta where it still existed in 1960. After direct involvement in the early phase of railway construction it retained a connection with the railways as a supplier of goods and services, e.g. bridge work and rolling stock from its Howrah Iron Works or loading and carrying contracts performed by another branch of the firm.

[8] Davidson, p. 147.

[9] Ibid., p. 148.

[10] Ibid., pp. 154–211, provides the best published description of the EIR line and its major works.

many contractors, including Nelson & Co. with 121 miles in Rajmahal, failed. The successes were the same three who had succeeded on the experimental line. The success of contractors in the North Western Provinces (the NWP Division of the EIR) was better, though Brandon & Co., a Cawnpore firm with the contract to build 27 miles in that vicinity, failed. Norris & Co., Hunt & Co. and Burn & Co. continued to be the main successes. Whenever a contractor failed, the EIR's engineers had to take over and, functioning as their own contractors, complete the line.

There were a number of reasons given by contemporaries for the failure of these men. Certainly the Santhal rebellion and the disturbances of 1857–8 played a role. The cost of labour and materials increased considerably—caused in part by the intensified demand generated by railway construction itself and partly by the disruptions. The transportation of certain materials from England to the worksites was slow; land assembly in the early going was delayed; government either interfered too much and slowed things down, or did not interfere sufficiently to help contractors overcome labour shortages by using local government authority 'to impart to the labouring classes that confidence which the contractors, left to themselves, found it hard to establish'.[11] The fundamental reason, though, was the deficiencies of the would be contractors: some were deficient in capital, many were deficient in experience and, perhaps, in motive. Another contemporary, admittedly one who advocated a system of petty contracts for the construction of the short Calcutta and South-Eastern line, referred to the 'vicious form' sometimes taken by the contract system on the EIR. Men who had capital (as some did) but 'neither practical knowledge of railway work nor plant "took contracts" with a view to reaping a large profit out of the difference of price received by them from the railway company and paid by them to the native contractors and labourers'.[12] In effect, it created a body of middlemen with little knowledge of or interest in the construction, while leaving the mobilization and management of the labour force in the hands of lower-

[11] 'The East Indian Railway', *Calcutta Review*, 56 (September 1858), pp. 241–2; also useful is 'Rajmahal, its Railways and Historical Associations', *Calcutta Review*, 71 (March 1861), pp. 110–43 and Davidson, p. 160. Hena Mukherjee, 'The Early History of the East Indian Railway, 1845–1879', (Ph.D. dissertation, University of London, SOAS, 1966) argues that difficulties with the assembly of land and materials slowed the early construction of the EIR. The author says very little about the mobilization of labour.

[12] IOL&R. Tract 530 (d), James A. Longridge, *Report on the Calcutta and South Eastern Railway, From Calcutta to the River Mutla, and the Extension Eastwards to Dacca and the Burmese Provinces* (London, 1857), p. 3.

level contractors/jobbers who would, in any case, be the ones, or even inferior to the ones, company engineers would employ under the petty contract system. A variant of this form of speculation appeared in western India where GIPR contracts were obtained by 'capitalists' and promptly sublet at reduced prices to a 'practical man' who had the knowledge and the experience to carry out the contract.[13] Ironically, the 'vicious' form of the large-contract system in the construction of the EIR may have been practised best by those who succeeded in completing their contracts. An established Calcutta firm like Burn and Co. would have had the knowledge and contacts to sub-contract the work of construction to petty Indian contractors.

Before turning to the GIPR it is instructive to consider another line in eastern India that was built in the late 50s and early 60s: the 110-mile EBR from Calcutta (Sealdah) to Kushtia on the Ganges. The contract for this line, dated 31 December 1858, was obtained by Brassey, Paxton, Wythes and Henfrey.[14] It was a complete, turn-key operation wherein the contractors agreed to finish the line in its entirety, supply a certain amount of rolling stock and maintain the line for one year after its completion for the sum of £1,045,000. Work began in 1859 and the line was opened for traffic in November 1862.[15] Here the vision of rapidly building an Indian railway through the instrument of a large-scale, British contracting 'firm' was realized.[16] Thomas Brassey (1805–70) was the largest of the large; he was undoubtedly the most successful railway contractor of his period. But his partners were no slouches: Sir Joseph Paxton of Crystal Palace fame; George Wythes, a prominent, wealthy railway contractor who took many contracts with Brassey and with others; and Charles Henfrey who, prior to going to Bengal as the resident partner and agent, had just built, as

[13] *Engineer's Journal*, 15 March, 1869, p. 47.

[14] The original signed contract and specifications still exists in IOL&R as the item in L/AG/46/10/2. Enclosed in a thin leather cover this 18″ high by 12″ wide document contains about 100 pages, each page covered by tight, clear hand-writing specifying, on a mile by mile basis, the details of the construction.

[15] Davidson, pp. 212–24 provides a description of the construction and the major difficulties that had to be overcome. Also see 'Eastern Bengal and Its Railways', *Calcutta Review*, 71 (March, 1861), pp. 158–84.

[16] 'Firm' needs to go in quotation marks because the British contractors did not form enduring, structured partnerships. Rather, two or more contractors would form a loose alliance based more on personal trust than on legal formalities in order to carry out a particular contract. Partnerships changed. In the same period Brassey partnered with Peto in one contract and with someone else to tender for another contract in competition against Peto.

Brassey's resident partner, the Ivrea Railway in Italy, and had been associated with him on other Italian lines.[17] Nonetheless, even this powerful and experienced combination found construction under Indian conditions difficult. As Henfrey remarked: 'Mr Brassey's usual good fortune did not attend him in this enterprise.'[18]

The partners lost money. The engineering difficulties were formidable but more crucial was the rise in the cost of labour—when it could be obtained at all—and materials, which affected the EIR construction as well. As Henfrey plaintively wrote to government: contractors had to assume risks but they were not speculative gamblers; they made every effort to identify costs before tendering for a contract but 'in the present instance, we contend that our losses have arisen from no want of due care and caution at the outset, but from an altered state of things in this country, which it was impossible for us to foresee'.[19] However, unlike the EIR contractors, Brassey, Paxton, Wythes and Henfrey had the resources and the will—a reputation to maintain which was important to their ability to win future contracts—to push on and bring the work to a successful conclusion, regardless of financial loss. Faced with such substantial and early losses, most firms, claimed the chief engineer, 'would have thrown up the Contract and forfeited their security. . .'.[20] The advocates of the use of substantial British contractors were correct in this instance.

The difficulties the partners faced and their proposed solutions take us to the heart of the on-going problems associated with mobilizing the resources, especially labour, needed to build railways in India in a timely fashion (given unlimited time, of course, railways could always be built, but this was rarely an option; there were usually time limits which, for contractors, were specified in their contracts). Henfrey found, first of all, that the partners needed a much larger superintending staff than their

[17] Helps, pp. 163–4, 181, 272–4. Another feature of these fluctuating partnerships is that the historian does not know the share of the risk assumed by each partner though clearly the loose partnerships were a device to spread the risk among a number of men. Brassey rarely took contracts by himself and he may have been a sleeping partner in some contracts. See Simmons, 'Introduction', p. xii. Brassey, Wythes and Paxton never visited India although the first two participated together, and separately with others, in a number of Indian railway contracts. Nor do we know the extent to which a given partner was involved in a particular contract beyond providing security and some capital. Indirect evidence suggests that George Wythes had limited involvement with his Indian contracts.

[18] Quoted in Helps, p. 273.

[19] IOL&R, L/PWD/3/62, Bengal RR Letters, no. 42, dated 25 June 1863.

[20] Ibid.

an experience had led them to expect and this increased their establishment costs from the 6–10 per cent typical in Europe, to 17 per cent in eastern Bengal. He also complained, as lesser points, of the slowness of government in furnishing land for the right-of-way (there being a need for detailed surveys and dealing with a multiplicity of owners and occupiers of land), of a prolonged and severe rainy season in 1861, and of difficulties in the transportation of plant and construction materials. He also alluded to the deleterious effects of government supervision.[21]

Recruiting and retaining labour was the central difficulty. In an effort 'to stem the tide of increasing rates' the firm made contracts with 'Native Sub-Contractors and others who well knew how to deal with the Coolies' and at first had some success until the sub-contractors had to abandon 'the Work with considerable loss'. Indigo planters were then tried since they presumably knew the district (Krishnagar) and commanded labour, but, despite offering the best rates the firm felt it could afford for earthwork, no planter accepted. 'During the first Working Season, the Coolies were but little desirous to work for us. They are at all times very shy and suspicious of strangers, besides which they were waiting to see what prices could be obtained by holding back.' Henfrey felt that had there been no competition for labour they might eventually have won 'the battle of rates', but they were surrounded by competitors: the EIR construction, the Calcutta and South-Eastern line, extensive Government public works on the Calcutta Circular Canal, the rectification of the course of the River Matabhanga, and works connected with the Calcutta Drainage which gave 'employment to a large number of workmen, who would otherwise have been available for the heavy Earthwork and Masonry about the Sealdah station'. And, in a sentence fraught with significance, Henfrey stated: 'The Native Sub-Contractors, Sardars and others were not slow to avail themselves of these heavy demands for labour.' Interestingly, the PWD engineer in charge of the Matabhanga works argued later that investigation had shown that the PWD Matabhanga earthwork rates were lower than those offered by Brassey & Co. but the coolies themselves earned better wages on the Matabhanga works because, under a petty contract system, a margin for the large-contractor's profits did not have

[21] Ibid. This material and the material in the subsequent two paragraphs, including quotations, comes from this RR Letter, unless otherwise cited. The particular document in question is Brassey, Paxton, Wythes & Co. to W. Purdon dated 19 February 1863 in which they set out their final accounts, explain why it cost so much more than estimated, and ask for some recompense. It is 19 pages of material to which Purdon's 13 page assessment, dated 7 March 1863, for the Agent of the EBR is attached. It is a useful, detailed document.

to be built in.[22] More money trickled into the hands of those actually doing the work or at least into the hands of the petty contractors who had the ability to obtain labour.

Finally, Henfrey inveighed against the system of advances, 'at all times the great curse of this country', to which the firm, despite its resolute opposition, had to capitulate. 'We soon found that if we did not follow the example of our neighbours, and tacitly sanction the system, we should get no Coolies, but such as resided in the villages in the immediate vicinity of the Line, who had not been previously accustomed to heavy works.' Once obtained, coolies were hard to retain because when the attempt was made to deduct part of the advances from their pay 'they would leave in a body during the night' and go to the competing works where they would get work and fresh advances 'leaving us no remedy against them but the slow, uncertain, and expensive process of Mofussil Law'.

The colonial state had weighed in on the side of capital.[23] Disturbances during the construction of the Bhore Ghat incline led to the Employers and Workmen (Disputes) Act (X) of 1860 which had a provision for fining or imprisoning workers who, having engaged to work for a particular period or to complete a particular task, refused to do so. However, the conditions of construction worker life and the uncertainties of law enforcement in the rural areas made Act X of 1860 and other remedies hard to enforce, as Henfrey found out. With the acquiescence of the colonial State, employers of railway construction labour had to find extra-economic and extra-legal ways to command labour.

Advances became, and have remained, a key element in the mobilization and retention of circulating labour in India.[24] The significance of advances for the worker for whom it represented the stake necessary to travel to a worksite—or even further, to tide him and his family over the unemployment of a rainy season—will be explored further in chapter four. However, advances helped to mobilize and tie labour to capital.[25] They subordinated the worker to the muccadum and, in the complex

[22] IOL&R, P/1195, PWD Progs., RR, May 1878, no. 44.

[23] Jan Breman, *Labour Migration and Rural Transformation in Colonial Asia* (Amsterdam: Free University Press, 1990), p. 61 argues that the official support for the subjugation of labour was one of the characteristics of peripheral capitalism in South Asia.

[24] Ibid., 'Seasonal Migration and Co-operative Capitalism', *JPS*, 6:1 (October 1978), pp. 41–70 provides a telling description of the role advances play in the current migration of labour to the sugar factories of Gujarat.

[25] The use of advances was not unique to India. See Stephen A. Marglin, 'What Do Bosses Do? The Origins and functions of Hierarchy in Capitalist Production',

hierarchy of supervision and direction that characterized Indian railway construction, the muccadum to the sub-contractor, the sub-contractor to the contractor (or to the engineer within the departmental system) and the whole hierarchy to the railway company or to the government that provided the capital. It must be emphasized that except for a few of the great British contractors, most of those who contracted to build railways in India had little capital of their own. This was still true in the early twentieth century when a railway official stated to a group of engineering students at Sibpur that some of the best contractors who could and would do good work had little capital. 'They have to borrow to commence work. It is most important these men should be kept paid for the work done well up to date and frequently. It helps them in their financial arrangements, and relieves their money pressure, which, if continued, is likely to result in bad or insufficient work.'[26] The acceptance by the British of the system of timely advances was one way in which construction became more expeditious. After some resistance the British came to understand better how the advance helped petty contractors obtain and retain labour and how it provided higher management with some control over the petty contractors.[27]

The manifold difficulties and obstacles identified by Henfrey were to re-occur in Indian railway construction, particularly in the earlier going, in various mixtures and with greater or lesser intensity. But particularly where obtaining, retaining and utilizing labour was concerned, Henfrey reveals much about the complex context of railway construction in nineteenth-century India. Railway construction meant the presence, however incomplete, of the forces (the labour process) and relationships (especially the commodification of labour power) of capitalism; railway construction penetrated, however temporarily in particular localities, deeply into the Indian countryside. The contractors and the engineers acting as contractors had the task of putting out capital. But to whom did they put

in A. Gorz (ed.), *The Division of Labour*, (Atlantic Highlands, NJ: Humanities Press, 1976), pp. 26–7.

[26] IOL&R, [300] A.78.F (c), Sir J.R. Wynne, *Notes on the Construction of Railways in India* (Calcutta, 1902), p. 7.

[27] Obtaining and retaining labour are also the key aspects in the development of a capitalistic labour market. H. C. Pentland put it well: 'The essential historical aspect of the capitalistic labour market, then, is the development of the supply and demand conditions that will support it. Two questions are especially critical: How are workers induced to flow into the labour market pool? And how are they prevented from flowing out again?' Pentland, 'The Development of a Capitalistic Labour Market in Canada', *Canadian Journal of Economics and Political Science*, 25:4 (1959), p. 450.

out the capital which they owned or had been entrusted with?[28] These men, even a partnership of the likes of Brassey, Paxton, Wythes and Henfrey, had to work through would-be petty capitalists, the layers of Indian intermediaries—sub-contractors and muccadums—who had the capacity to mobilize labour but who were themselves only partially enmeshed in the values and relationships of the capitalist order. These lower-level intermediaries, in turn, had to recruit labour from peasant and tribal societies whose linkages to the emerging world of capitalist relationships in the third quarter of the nineteenth century were limited in many dimensions: economic, social and cultural.[29] Free labour had a limited presence in the Indian countryside in the 1850s and 60s although, partly as the result of railway construction, it grew as the decades passed.[30] Even so, labour never became an atomistic entity, bargaining directly, freely and solely over wages with would-be employers. Extra-economic links bound workers to muccadums and muccadums to sub-contractors. It was these links rather than legal sanctions that best enforced the provision of labour power in repayment of the advance. It was a compliance, the British learned, best extracted by Indians. The British managed the construction process: they usually (and especially in earthwork) did not directly command the labour of Indians.

Henfrey's complaint about the slow uncertainty of Mofussil Law was one example of the British inability to enforce compliance with the advance. Another telling example, both with respect to the problem and the solution, comes from the construction of the Bengal–Nagpur line in the Pendraghat area in 1890.[31] The ghat section was being built departmentally and tunnelling was involved. To commence, the engineer imported 30 men who had worked on the tunnels of the southern Mahratta Railway. They were provided with advances and train fares but soon after

[28] Part of the capital, of course, went into fixed assets (plant and rails) often obtained from Britain and which represented some of the backward-linkages that benefitted Britain's industrial economy.

[29] Relevant here is Jacques Pouchepadass, 'The Market for Agricultural Labour in Colonial North Bihar 1860–1920', in M. Holmstrom (ed.), *Work for Wages in South Asia*, (New Delhi: Manohar, 1990), pp. 10–27.

[30] The recruitment and organization of labour for India's coal industry, whose nineteenth-century development was spurred by the fuel needs of the railways, presents interesting similarities and differences to the case of railway construction labour. See C.P. Simmons, 'Recruiting and Organizing an Industrial Labour Force in Colonial India: The Case of the Coal Mining Industry, c. 1880–1939', *IESHR*, 13:4 (October–December 1976), pp. 455–85.

[31] Hertford Record Office, no. 86726, Leake Papers: 'Report, dated 20th February, 1890, on the Pendraghat District by Mr. S.M. Leake, District Engineer.'

their arrival they began to run away and none were left after two weeks.[32] The British engineers abandoned their effort at direct recruitment. They subsequently obtained the needed labour 'through mukadums or mates, who make advances to the men, to whom we give high pay in order to cover losses caused them by men leaving the work and failing to make good the advances received'.[33] Elsewhere in the district earthwork was being done by contract and labour supplies were good: 'All the coolies worked under advances, each contractor keeping men out in the parts of the country where labour was obtainable, collecting and advancing money to the men. This method of working has been fairly successful and the contractors have lost very small sums compared to the large advances they have had to make.'[34]

What this material reveals is, firstly, the layering of intermediaries and the difficulties the British encountered when they tried to contact labour directly. Secondly, it shows a mix of systems of construction, departmental and contract, both involving advances. Thirdly, it shows how clever managers could turn mates into waged recruiters by substituting high wages for advances but making it clear that those wages were meant to cover losses the mates might encounter as the result of absconding coolies—in short the wages were a form of advance but the risk was shifted to the mate. The lesson for the British was clear: they had to give advances to those they could trust and over whom they had some control.[35]

Individuals who were free to sell their labour power to the Brassey partnership were few and far between. The emphasis in Henfrey's lament is on groups of workers who came and left together, often, one suspects, at the command of native sub-contractors, sardars and others who 'were not slow to avail themselves of these heavy demands for labour'. As Bayly has observed, the changes to India's political economy developing coterminously with the extension of British rule did not initiate a free market

[32] Note that they did not leave collectively at the same time. This suggests the advances in this case may have gone to the individual workers. Tunnelling specialists were few in number.

[33] Leake Papers, 'Report', p. 2.

[34] Ibid., pp. 2–3. Among the coolies were people from Chhattisgarh who preferred to work for ten days, return to their village to obtain another ten days supply of rice, etc. and then return again to the works for another ten days and so on, in the belief that this was more financially rewarding than staying on the works and buying supplies at the ghat where prices were high.

[35] The later century manuals for the engineers stress the need to be cautious when hiring petty contractors.

or undermine the headman system. Rather, all rights and perquisites came up for sale but 'what made the sale worthwhile was that *within* the "little dominion" competition and the free market were still excluded. Here, political muscle, the authority of the headman and the rights of caste rank continued to operate to produce cash, labour or commodities'.[36] Railway construction needed lots of waged labour but that labour had to be tapped primarily through various headmen who, within the ambit of capitalist relationships, functioned as sub-contractors or even sub-sub-contractors. In the case of local, daily-hired labour, the 'headman' may have been a village power-holder whose goodwill was purchased in order to obtain the temporary release of village labour power.

Put another way, railway construction took place within a social formation characterized by the presence of capitalist and pre-capitalist forces and relationships. Henfrey, his contemporaries and successors were confronted with the problem of linking down and across these different structures, if they were to get the labour they needed. Their problem was magnified by the fact that not only did they come from the capitalist world, but that that world was white, Victorian Britain, thousands of miles away. Colonialism widened distances at the level of social relationships because it created patterns of domination involving socio-culturally different entities.

Meanwhile, on the West Coast, the GIPR was actively pursuing the construction of its lines with a similar commitment to the large-contract system, although seemingly with greater success and with the precaution of making the initial contracts for quite short lengths of line which, nonetheless, did involve, in some cases, heavy and difficult work.[37] The 33 miles from Bombay to Kalyan, at which point the line divided to go north-east for 26 miles to the Thal Ghat and south-east for 38 miles to the foot of the Bhore Ghat, was divided into seven contracts with the

[36] C. A. Bayly, *Rulers, Townsmen and Bazaars. North Indian Society in the Age of British Expansion, 1770–1870* (Cambridge: CUP, 1983), p. 317. Emphasis in the original. The significance of Bayly's comment was brought home to me by Dipesh Chakrabarty, *Rethinking Working-Class History* (Princeton: PUP, 1989), pp. 112–13, who argues that the sardar's control of labour lay heavily in the pre-capitalist relationships of community, kinship, religion and their accompanying norms and ideas but that the sardari system proved adaptable to the needs of labour mobilization in the jute industry.

[37] Berkley, *MPICE*, 19 (1859–60), pp. 586–624 provides a useful description. Berkley was the CE and on pp. 606–7 he praises, without reservation, the large-contract system.

building of workshops, stations and bungalows, etc. being tendered as separate, smaller works.[38] The difficult climbs at the Bhore Ghat for the south-eastern branch of the GIPR and at the Thal Ghat for the north-eastern branch were let as two large separate contracts.

The demanding 15 miles up the Bhore Ghat, contract no. 8, was first taken and then given up by William Frederick Faviell (1822–1902).[39] Faviell, who later built railways in Ceylon, was in India to make, not to lose, money. It is clear that throughout his activities in India, Faviell sought to maximize his own return to the detriment of those, Indian and European, who worked for him. His European agency on the Bhore was never fully adequate nor very savoury and he had little concern for his Indian workers: he paid low wages, provided inadequate housing, and then complained about labour shortages. His junior partner on contract no. 1 called him a 'conceited and obstinate man' with a 'very disagreeable and offensive manner' who treated people under him 'as if they were dogs or inferior beings'.[40] The Bhore Ghat contract was taken over by Solomon Tredwell, another well-known English contractor, who reached Bombay on 29 October 1859, and died on 30 November 1859 from a disease acquired on the Ghat. The widow, Alice Tredwell, then took over the contract and, with the help of former GIPR engineers who became her resident engineers, eventually saw the contract through to its conclusion.[41]

The extensive mileage of the north-east and south-east lines of the GIPR above the Ghats were let out on a series of large contracts and here failures became more common. Duckett and Stead failed on their contract no. 15 for the 138 miles of the line from Bhusawal to Harda, despite W. Stead stepping quickly into on-site supervision of the contract after

[38] IOL&R ST 1862, *First and Second Annual Reports of the Great Indian Peninsula Railway Company, for the Years 1854–55—1855–56* (Bombay: Education Society's Press, 1857) and *Report to the Shareholders of the Great Indian Peninsula Railway Company on the Progress of the Works in India* (London: W. Lewis & Son, 1857) provide information on these and subsequent contracts.

[39] Obituary, 'William Frederick Faviell', *MPICE*, 150 (1902), pp. 463–6.

[40] IOL, Mss. Eur. C. 401, Fowler to Leather dated 2 May 1851. An employee of Faviell, Godfrey Oates Mann, who was in Bombay at the same time wrote accounts of his employer which generally confirm Fowler's statements. See the letters in IOL&R, photo. Eur. 197.

[41] Tredwell's death and the subsequent arrangements can be followed in MSA, PWD (Railway) 1859, vol. 25, compilation 447 and 1862, vol. 6, compilation 317. The engineers in question, Adamson and Clowser, later took on their own account the contract to build the GIPR line from Sholapur to Kulburgah.

his father, Abraham Stead, died at Asirgarh on 7 November 1859.[42] George Wythes and Co. took over the contract in March 1862 and eventually brought it to completion, though not without frequent complaints about their slow progress.[43] Joseph Bray, Wythes and Jackson, Hood, Winton and Mills, Lee, Watson and Aiton, and Norris and Weller were among others who took substantial contracts with greater or lesser degrees of success.[44] The partners, Norris and Weller, 'two very upright but two very old and incapable colonels' who got 'to play for a while "at contracting" ', took over one of the contracts (no. 17) of the failed Duckett and Stead and were, in turn, conspicuous failures.[45]

Apart from the turnover among the contractors and sub-contractors as they gave up or had their contracts resumed, it is also clear that even when the initial contractor completed his contract, the work was often done slowly, cheaply and shoddily. In February 1859, for example, a government engineer reported after an inspection of contract no. 12 (Jackson and Wythes 72 miles, Chalisgaon to Bhusawal) that it had been cut-up and sub-let to English contractors who were 'neither educated nor respectable nor scrupulous'.[46] If the arguments advanced above in the discussion of the EBR are correct, these European sub-contractors represented another layer on top of those Indian 'headmen' who commanded labour. Furthermore, the presence of extensive sub-contracting to Europeans suggests that some of the takers of large contracts were little more than financial speculators.[47] This supposition is supported by a report in

[42] MSA, PWD (Railway) 1859, vol 55, compilation no. 433: Report on the Death of Mr Stead, Contractor on the Line between Bhusawal and Jubbulpore. The lenient treatment of Faviell at the time of his giving-up the Bhore Ghat contract made it difficult to deal harshly with Duckett and Stead. See MSA, PWD (Railway), 1859. vol. 30, compilation 460: Inability of Duckett and Stead to Complete Contracts Nos. 15 and 17. It is not clear if they ever began no. 17 (117 miles, Sohagpur to Jubbulpore) which came into the hands of Norris and Weller.

[43] IOL&R, PWD RR Progs, August 1864, nos. 12–15.

[44] *Herapath's Railway Journal*, 26 April 1862, p. 438.

[45] The characterization of their abilities comes from *The Bombay Builder*, Vol. III (5 September 1867), p. 96. Complaints about their work on contract no. 17, the 117 miles from Sohagpur to Jubbulpore, abounded prior to the Company taking over. For example, see MSA, PWD (Railway), 1863, vol. 22, compilation no. 94.

[46] MSA, PWD (Railway) 1859, vol. 45, compilation no. 108: North Eastern Extension. Capt. Rivers Report on the General State of the Works from Wassind to Bhusawal.

[47] This is also suggested in 'The Contract System', *The Bombay Builder*, vol. III (5 September 1867), pp. 92–9.

1867 which condemned Wythes and Jackson's conducting of contract no. 15 (north of Bhusawal) as dilatory and unsatisfactory. Management lacked vigour, a parsimonious attitude prevailed, and even ordinary tools, let alone larger pieces of machinery, were conspicuously absent.[48]

This suggests that these major contractors and their sub-contractors put little of their own capital into the project and concentrated on work that could be performed by manual labour, paid for by advances from the GIPR. C.B. Ker, a former CE of the GIPR, was the chief agent for the contractors. One suspects, he probably functioned more as a major sub-contractor and made his share from the difference between his sub-contract 'tender' and his expenditures on the actual construction. As for shoddy work, investigations in the later 1860s showed that on the GIPR, some 2000 buildings, bridges and other masonry structures had either failed or needed extensive repairs.[49]

Charges that some contractors had 'scamped' their work were still current in the early 1880s, though one contractor, Thomas Glover, who took some of the rebuilding contracts, including the large Mhow-ke-Mullee viaduct on the Bhore Ghat, argued that the fault lay in the imperfect knowledge of Indian building materials possessed by British engineers in the early 1860s.[50] Incidentally, the Mhow-ke-Mullee viaduct, the largest on the incline located immediately below one of the longest tunnels, collapsed on 19 July 1867. The viaduct had been inspected in the early morning. A train with passengers passed over at 6:50 a.m. A goods train arrived near the viaduct at 7:40 a.m. when, at a distance of 200 yards, 'the Native Fireman, who was on the lookout on the outside of the curve, gave the alarm that the bridge was falling, the train was immediately stopped, and the Driver and he saw the remaining arches go'.[51] The fireman crossed the ravine on foot to warn a goods train known

[48] IOL&R, L/PWD/3/69, Bengal RR Letters, no. 12 dated 12 February 1869. The author of the report was the CE of the Central Provinces. An obituary of George Wythes (1811–83) states: 'He undertook the early and more difficult part of the work of the Great Indian Peninsula Railway, *from which he derived considerable profit.*' Obituary, 'George Wythes', *MPICE*, 74 (1883), p. 295. The emphasis is mine.

[49] *PP (Commons), 1872*, cmnd 327, 23 July 1872, *Report from the Select Committee on East India Finance*, para 1895.

[50] *PP (Commons), 1884*, cmnd 284, 18 July 1884, *Report from the Select Committee on East India Railway Communication*, paras 2065–9. The charge of scamping was made by a member of the committee, while putting a question to Glover.

[51] IOL&R, L/PWD/3/66, Bengal RR Letters, no. 108 dated 24 October 1867. This description from the records can be compared with a somewhat more 'dramatic'

to be descending, which he successfully did just before it entered the tunnel. A double catastrophe had been averted.

Most interesting, and among the most successful, of the early GIPR contractors was Jamsetji Dorabji who satisfactorily, expeditiously and at less cost than his European competitors, completed at least five important contracts.[52] He stands out in the early decades as an Indian who successfully conducted large-scale contracts. Of course, Indians played crucial roles as sub-contractors, but Jamsetji Dorabji succeeded at the higher level. Why? One can only speculate, but perhaps his position among the Parsis, that most westernized of the mid-mineteenth-century Indian communities, had something to do with his ability to seize the opportunities presented by railway construction. Although Parsis became an important trading and financing community, they emerged as such from an artisan background in western India in ship-building and by acting as brokers for European traders.[53]

Jamsetji Dorabji Naegamwalla was born in Surat in 1804 and died in Bombay in 1882.[54] He left home at the age of 12 to enter the service of a Parsi master-builder and then was employed for four years as a carpenter in the Government Dockyard, Bombay. He began to find patrons and acquaintances among the European community and, finding that he possessed the ability to command labour, began to take small contracts. His first contract was to construct salt-pans near Sweree which belonged to Dadabhoy Pestonji Wadia. Further contracts followed from other Parsis and then from the Bombay Municipality. He tendered for GIPR contracts no. 1 and no. 2, but failed to get them. He tendered the lowest bid for contract no. 3 which, though not supported by CE Berkley, was given to him by the GIPR's Committee of Direction because the four

---

account presented in J.N.Sahni, *Indian Railways One Hundred Years, 1853 to 1953* (New Delhi, 1953), p. 48. Davidson, p. 275 notes that many viaducts on the GIPR failed. 'Failure was caused either by inefficient foundations, bad dry mortar, neglect of bond, or at times by the unequal settlement of a hard casing of ashlar or block in course, and of a soft hearting of rubble. There was also much laxity in supervision.'

[52] Berkley, *MPICE*, 19 (1859–60), p. 607.

[53] D.R. Gadgil, *Origins of the Modern Indian Business Class. An Interim Report* (New York: Institute of Pacific Relations, 1959), p. 20.

[54] Brief biographies of Jamsetji can be found in H.D. Darukhanawala, *Parsi Lustre on Indian Soil*, vol. I (Bombay, 1939), pp. 198–9 and Dosabhai Framji Karaka, *History of the Parsis including their Manners, Customs, Religion, and Present Position*, vol. II (London: Macmillan and Co., 1884), pp. 253–7. I am indebted to Frank Conlon for these sources and to his generous provision of extracts from them.

European tenders were 'exorbitant' and it was known that Jamsetji 'had executed other considerable works by contract, and that he was a man of energy and resources. . .'.[55]

The successful completion of contract no. 3 established Jamsetji as a railway contractor, but at one stage it seemed likely he would fail. Great loss and trouble loomed as the inexperienced Jamsetji struggled with a deep, wet bridge foundation as the monsoon rains approached. Fortunately for Jamsetji, a GIPR assistant engineer named Clowser provided assistance. The Parsi contractor pulled through and went on to win and complete additional major railway contracts.[56] By May 1855 he had earned Berkely's accolade as 'the foremost Native Railway Contractor in India' who commanded the means and the men to assist greatly in the prosecution of railway and other works, and whose efforts had created a positive name for railway construction 'throughout the remote sources of labour and material, upon which we so greatly depend. . .'.[57]

Indeed, Jamsetji's ability to obtain and retain labour was the key to his success as a railway contractor. His European rivals on nearby contracts, e.g. Faviell on the Bhore Ghat and Wythes and Jackson on the Thal Ghat, claimed they could not obtain and retain the requisite supply of skilled and unskilled labour, yet they admitted that Jamsetji hired workers away from them.[58] Faviell solved his problem by relinquishing his contract; Wythes and Jackson, in one instance at least (the Thal Ghat incline), pursued a comparatively rare solution in Indian railway building, namely a more capital-intensive approach involving the greater use of mechanical appliances such as an elaborate system of temporary wagon-tramways by which earth was tipped into some of the large embankments.[59] Perhaps

[55] *The Bombay Times Overland Summary of Intelligence*, from 1–11 May 1855, pp. 70–1. This material, and what follows, comes from speeches at a celebration on 30 April when Jamsetji invited the leading citizens of Bombay to visit his virtually completed Kalyan to Thal ghat contract (no. 10) and to enjoy a sumptuous banquet and entertainment afterwards in a 'suite of tents on a scale of the utmost liberality and splendour' laid out adjacent to his 12 arch Kalloo viaduct. The whole affair represented Jamsetji's affirmation of his success and the illustrious guest list reinforced the fact that this was a Parsi who had succeeded in a European venture.

[56] Ibid. Jamsetji subsequently claimed he had 'suffered pecuniary loss' on contract no. 3. Bombay RR Letters, no. 76 dated 12 December 1861. Clowser was to remain involved in GIPR construction in a number of important capacities through the 1860s.

[57] *Bombay Times*, 1–11 May 1855, pp. 70–1.

[58] MSA, PWD (Railway), 1859, compilation 206.

[59] James J. Berkley, *Paper on the Thul Ghaut Railway Incline* (Bombay: Education

Jamsetji enjoyed more success in his efforts to obtain and retain labour because, as one biographer stressed, he was 'an excellent master to his workmen, not only liberal in wages, but pleasant of speech, and constantly throwing himself into contact with them, it is not at all extraordinary that native labourers should have taken employment with him, even in the worst districts for the supply of water and unhealthiness of climate'.[60] This portrait, allowing for the sympathies of a biographer, is in marked contrast to what was said about Faviell and some other European contractors and sub-contractors. Moreover, it suggests something of the continuing importance of personal qualities and relationships within a capitalistic labour market; the more benign of the extra-economic dimensions repeatedly mentioned in this study.

There were additional reasons for Jamsetji's success. First, he was a man of considerable talent, whose powerful and retentive memory compensated for the fact that he was illiterate. He was particularly adept at arriving at accurate estimates upon which to base his tenders—a crucial task for any contractor who expected to make a profit.[61] Second, he learned from experience and knew the value of employing experienced subordinates. The same Clowser who saved Jamsetji on contract no. 3 was hired as his manager for contract no. 10. Jamsetji acknowledged he had learned the value of the European way of executing work from his observation of Clowser's management of contract no. 10, and stated that he intended to turn that knowledge to good account: an acknowledgement that supports my argument about the importance of management within the railway construction process.[62] Third, Jamsetji appears to have had financial backing in the form of partners or supporters (investors?) in the Parsi community to whom he could turn for financing or security. Certainly his membership in a westernizing community, among whom there were men of wealth, including other contractors, enhanced his opportunities for success.[63] He refers to the death of two partners in the course of contract no. 3; at the same time, a request that he obtain two approved people to provide security before he received the go ahead to rebuild a collapsed viaduct on contract no. 10 was met with an indignant,

Society's Press, 1861), pp. 24–7. Given Wythes' parsimony on later contracts, it was a rare solution for him also.

[60] Karaka, Vol. II, p. 256 quoting an unnamed European biographer.

[61] Ibid.

[62] *Bombay Times*, 1–11 May 1855, pp. 70–1.

[63] Berkley mentioned six other Indian contractors in his speech at Jamsetji's 30 April celebration, of whom at least three were Parsis.

'it brings disgrace to my character'. (Jamsetji denied responsibility for the collapse; he had followed the design and specifications. The dispute was to go to arbitration, but in the meantime the viaduct had to be rebuilt, at the cost of 2 lakhs of rupees, so that the line could reopen.)[64] The friends or relatives he might call upon to provide an endorsement would think that if the GIPR, for whom he had satisfactorily carried out contracts for nine years, had such little confidence in him as to not accept his personal security, then 'how could they, and the consequence would be detrimental to my credit, and I feel it repugnant to do so. . .'.[65] He was permitted to go ahead on his own security. He never took another contract from the GIPR although he did subsequently build a large portion of the BB&CIR beyond Ahmadabad.[66] Jamsetji retired in the early 1870s, in his mid-sixties, wealthy and well-known, and died at Malabar hill, Bombay, on 18 September 1882.[67]

The SP&DR also tried the large-contract system. The British contractors, James and Edwin Bray, were awarded the job of building the 105 miles from Karachi to Kotri.[68] They began work on their contract in March 1858 and abandoned the contract in June 1859. They must have been under-capitalized because they lacked an adequate plant and proved unable to pay their workmen on a regular basis.[69] The Railway Company then took up the construction departmentally, that is through the Company's engineers acting as their own contractors. Wage rates, 25 per cent below those offered by the Brays, were established by the engineers but daily payments of wages were guaranteed. 'This had the effect of drawing

[64] For the mention of partners see the *Bombay Times*, fn. 62. For contract no. 10, IOL&R, L/PWD/3/271, Bombay RR Letters, no. 76 dated 12 December 1861. Presumably Jamsetji dictated the letter.

[65] Bombay RR Letters, no. 76

[66] Karaka, vol. II, p. 255. Perhaps the episode soured his relations with the GIPR but more likely he carried through on his announced intention to retire 'for a while to recruit my strength'. Bombay RR Letters, no. 76.

[67] *The Times of India*, 20 September 1882, p. 4.

[68] Brief descriptions of the building of the line in Sind can be found in H.C. Hughes, 'The Scinde Railway', *Journal of Transport History*, 4 (November 1962), pp. 219–25 and P.S.A. Berridge, *Couplings to the Khyber* (New York, 1969), pp. 24–34. One of the Bray brothers obtained in 1856, a favourable contract to build the Vihar Water Works to supply good water to Bombay. See Mariam Dossal, 'Henry Conybeare and the Politics of Centralised Water Supply in Mid-Nineteenth Century Bombay', *IESHR*, 25:1 (January–March 1988), pp. 91–4.

[69] Brunton, *MPICE*, 12 (1862–3), p. 455.

a considerable number of labourers to the undertaking.'[70] The construction proceeded to a successful conclusion and the line was formally opened in May 1861. The Brays in the meantime launched extensive claims for recompense which went to an arbitration whose proceedings ran on for years.[71] Some 5700 individual items were in dispute and, as John Brunton, the CE of the Sind Railway, recalled, his testimony in the case extended over two years and the cross-examination over nine months.[72] This was by no means the last time that a railway company or the Government of India found itself embroiled in a long arbitration or court case with a railway contractor.

To the north in the Punjab, the same Company undertook to build a line from Lahore, 219 miles south to Multan and 32 miles east to Amritsar. Work began in early 1858. The formal opening of the short stretch to Amritsar took place in April 1862 and of the Multan section in April 1865.[73] Again the initial orientation of the Company was to work through substantial contractors. But how substantial and from where? The search for an answer to these questions involved some debate and the implementation of the first set of answers led to controversy and failure.

British officials in the Punjab, led by Chief Commissioner (soon to be styled Lieutenant-Governor) John Lawrence, argued for the use of Indian contractors and against the importation of contractors from England. Lawrence felt that if European contractors with sufficient capital were already on the spot and available they were to be preferred but

> such persons are rare; especially in the Punjab while substantial native contractors are ready to come forward to any required extent, and these men, having an intimate acquaintance with all the resources of the country; and many of them having already acquired, as contractors in the Canal and other Engineering

[70] Ibid.

[71] IOL&R, L/PWD/5/4/18. Compilations and Miscellaneous: 'Bray v. Scinde Railway Company. Arbitration 1862–1868'.

[72] John Brunton, *John Brunton's Book, being the Memories of John Brunton, Engineer, from a Manuscript in his Own Hand Written for his Grandchildren and Now Printed.* With an introduction by J.H. Clapham (Cambridge: CUP, 1939), p. 154. After all his efforts Brunton was disappointed to find the case settled by what he thought was 'a fearful blunder—a sort of compromise amongst the lawyers, which astonished the Arbitrator as much as it did me'.

[73] Davidson, pp. 310–14 provides general coverage but also see Berridge, pp. 35–40 and G.S. Khosla, 'The Growth of the Railway System in the Punjab', in Harbans Singh and N. Gerald Barrier (ed.), *Essays in Honour of Dr. Ganda Singh* (Patiala: Punjabi University, 1976), pp. 283–90.

departments, experience in works analogous to those of the railways, must obviously possess immense advantages over persons arriving from Europe, with no previously acquired Indian experience.[74]

Lawrence went on to make the point stressed earlier in this chapter. Below a certain level in the construction process, management had to be in the hands of Indians: they commanded or knew how to command labour and had to be involved. Large European contractors would simply be an extra layer of go-betweens, so why not have the engineers deal directly with Indian contractors. Indians also often commanded supplies, such as timber or material for ballast, vital to the construction.

John Lawrence's opinion carried a great deal of weight in 1858. The 'Saviour of the Punjab' and his band of like-minded subordinates had successfully weathered the events of 1857–8. They had prestige and they could claim intimate knowledge of local conditions. It is not surprising that we find William Brunton, CE for the SP&DR in its Punjab section, writing to his Board of Directors that there were plenty of native contractors available who had executed large works on roads and canals and who would be delighted to take on 30–40 miles of line exclusive of the permanent way.[75]

At some point in the tendering process, the idea of a number of Indians taking 30–40 miles each ended with two men taking the lion's share of the work. Jumna Das of Multan had the contract for some 110 miles of earthwork northward from that ancient city, and Muhammad Sultan of Lahore eventually ended up with much of the rest, encompassing some 142 miles of earthwork, the Lahore and Amritsar station buildings, plus large contracts for the supply of sleepers and ballast. Others obtained lesser contracts: Omerdeen of Lahore got a contract to build bungalows and wells; Mr. Coates of Ferozepore got the important task of transporting permanent way materials from Karachi to the Punjab; Ter Arratoon, an Armenian living in Lahore, got a sleeper contract; and Mela Ram, a contract to supply telegraph posts.[76] A 'Chevalier De Cortanze' later was

[74] IOL&R, L/PWD/3/56, Bengal RR Letters, no. 1 of 1858, enclosures. Offg. Sect. to the Chief Commissioner to Sect. to Govt of India, 21 October 1857

[75] IOL&R, L/PWD/3/56, Bengal RR Letters, no. 1 of 1858, enclosures. Brunton to Directors, 16 June 1857. William was a brother of John Brunton in Sind. The SP&DR, though one company, was divided into separate divisions: Sind, Punjab and Delhi plus the Indus River Flotilla that provide a river-born connection between Multan and Kotri.

[76] IOL&R, L/PWD/3/59, Bengal RR Letters, 1860, enclosures. Agent, Punjab Railway to the Chairman of the Scinde Raialway Company, no. 585, dated 31 October 1860.

to get contracts for the provision of ballast and also for the construction of 27 miles of line east of Amritsar, until the entire Amritsar–Delhi line was put in other hands.[77] Chota Lal, Deonath, Hookum Singh and Prem Singh also received ballast or sleeper contracts somewhat later.[78]

Jumna Das, Omerdeen and, most important, Muhammad Sultan, eventually failed. Sultan, about whom something is known, provides an instructive comparison with Jamsetjee Dorabjee with whom he stands as one of the few Indians who took large contracts in the early period of India's railway development.

Muhammad Sultan (c. 1809–75), a Kashmiri Muslim resident in Lahore, was a soap manufacturer and a wrestler during the days of Ranjit Singh.[79] When the British annexed the Punjab he grasped the opportunities opened up by the British public works programs. He provided bricks from old buildings for the construction of Lahore's Mian Mir cantonment in 1850.[80] By the mid-1850s the British were referring to him as an 'enterprising contractor' who had made 'large advances to the villagers, for the purposes of procuring carts of good quality, and he is able to meet any demands for this description of carriage which may be made'.[81] Then, with his credentials as a PWD contractor and his links with British officials established, he moved extensively into railway contracting—indeed he sought the contract for the entire length of the Lahore–Multan line.

By 1860, Sultan was the major contractor in the central Punjab. Wealthy, honoured by the British in a public ceremony and known for his liberal charity (he had, among other things, built a serai in Lahore for the poor, just outside the old city's Delhi Gate), he was appointed to Lahore's first municipal committee in 1862. But at the moment of his

[77] IOL&R, L/PWD/3/61, Bengal RR Letters, no. 50 of 1862, enclosures. The same material tells us of the failure of Jumna Das.

[78] IOL&R, L/PWD/3/62, Bengal RR Letters, no. 79 of 1863, enclosures.

[79] Syad Muhammad Latif, *Lahore* (Lahore: New Imperial Press, 1892), pp. 169–70.

[80] NAI, Govt of India, Foreign Dept, Political Progs, nos. 88–92 of 6 September 1850.

[81] R.H. Davies, R.E. Egerton, R. Temple and J.H. Morris, *Report on the Revised Settlement of the Lahore District in the Lahore Division* (Lahore, 1860), p. 4. Later, Sultan was to raze the mosque of Sitara Begam for its bricks. The mosque had been used as a residence for the English editor of the *Lahore Chronicle* and then became the property of some railway officials who sold it to Sultan. Latif, p. 170, states that Lahore's Moslems attributed Sultan's subsequent misfortunes to his destruction of this house of God.

greatest success Sultan was already in difficulties. Complaints, small and muted at first, began to be made about deficiencies in his performance. Ballast was not being supplied in sufficient quantity and he was well behind on his sleeper deliveries. The complaints mounted, and Sultan lost many of his contracts. This, partly through government intervention, was modified and Sultan signed an agreement in May 1861, relinquishing all construction work on the 110 miles of line south from Lahore. He retained his contracts to supply sleepers, some dressing and ballast, and to build the Lahore station works and the embankment leading to the Amritsar station.[82] De Cortanze got some of the work, Coates others, including the Amritsar station buildings. In 1863 the Railway Company's Agent labelled Sultan an 'effete and quasi-bankrupt Contractor' who, nonetheless, held a near monopoly on two urgent necessities of the railway, timber and ballast, but who did not have the 'energy or the means for properly working either'. Sultan, moreover, still had PWD contracts and retained 'the consideration from the Government'.[83] Soon his railway contracts disappeared. By 1870 Sultan was deeply in debt and his many properties mortgaged to his major creditor, the Maharaja of Kashmir, with whom he had become involved in his search for supplies of timber.[84] He died in straitened circumstances, in 1875.

The reasons Muhammad Sultan failed were no different from those that explain the failures of English contractors, whether from England or residing in India: lack of capital and/or over-extension, lack of organizational skills and a cost-price squeeze generated by the increase in labour rates or the costs of materials. Indeed, all these factors were advanced in Sultan's defence by no less a figure than John Lawrence. Sultan had limited capital and operated heavily on credit; he took extensive railway contracts 'of, to him, an altogether novel and complicated description'; his costs had increased after his tenders were accepted, and so on.[85] Sultan,

[82] IOL&R, P/190/64, PWD Progs., RR, nos. 23–6 of 6 August 1861.

[83] IOL&R, L/PWD/3/62, Bengal RR Letters, no. 13 of 1863, enclosures. The continuation of Government patronage was remarkable. On 26 January 1867, the *Lahore Chronicle*, p. 79 noted that Sultan had received the contract to build cutcherry buildings in Lahore in preference to Mela Ram which 'we think but fair; for the former has been for years employed by Government, and has given considerable satisfaction both with regard to the completion of his work and his unbounded liberality'.

[84] *Indian Public Opinion and Punjab Times*, 20 December 1870, p. 9021. Another connection was the fact that his brother, Abdul Rahman, was in the Maharaja's service. See *Indian Public Opinion*, 8 March 1872, p. 548.

[85] IOL&R, P/190/64, PWD Progs., RR, no. 24 of 6 August 1861.

however, was enmeshed in a set of social relationships in Lahore that denied him the luxury of simply abandoning his contracts, unless 'his name and credit would not suffer'.[86] In this he shared the concern that Jamsetji Dorabji exhibited at the time the latter was asked to provide independent security for the rebuilding of a failed viaduct. Indian contractors had concerns of status and prestige—'izzat'—that did not trouble many English contractors, although, interestingly, at the other end of the scale a contractor like Brassey and his partners did try to maintain their reputation as men who completed their contracts—a form of 'izzat'. In terms of capital resources, however, Brassey and Sultan were far apart.

It is easier to account for Sultan's eventual failure than to explain his considerable success, although the fact that he was once a wrestler offers a tantalizing hint. If he was a good wrestler he would have been well known.[87] There has been, in the history of Indian labour, a continuing connection of the jobber (the sirdar, the muccadum) with the gymnasium and with physical prowess.[88] Regardless of that we may see in Sultan's eventual difficulties, the same qualities of enterprise, vigour and acumen that had led him first to grasp the opportunities presented by the British PWD programme and then, encouraged by his first successes, to over-reach himself because 'being quite illiterate, the vast extent of the contract he has taken from the Railway Company appears to have succeeded his powers of superintendence. . .'.[89] Illiteracy was a handicap (one that Jamsetji overcame) in an endeavour involving the widespread mobilization of men and material and the accounting thereof. However, the psychological speculations that inform some of the literature on entrepreneurship are rarely amenable to historical reconstruction; we will never know precisely why Sultan put capital and credit at risk.

Nonetheless, it is tantalizing to note that he began his contracting career by becoming, in part at least, a client of British officials who acted as his patrons and who never abandoned him. The newly-annexed Punjab of the early 1850s was run on paternalistic lines. Railway construction in the mid and late 1850s involved the more intensive intrusion of the

[86] Ibid., Capt. E. G. Sim, Under Secretary to the Punjab Govt, in the Railway Dept. made this observation.

[87] Wrestling was and is a popular sport in north India. See Latif, p.266 and Nita Kumar, *The Artisans of Banaras. Popular Culture and Identity, 1880–1986* (Princeton: PUP, 1988), esp. pp. 111–24.

[88] Chakrabarty, pp. 109–12; Dick Kooiman, *Bombay Textile Labour. Managers, Trade Unionists and Officials 1918–1939* (New Delhi: Manohar, 1989), p. 22.

[89] IOL&R, L/PWD/3/61, Bengal RR Letters, no. 50 of 1862, enclosures.

practices of capitalism, with tight contracts, specifications, expectations of performance and notions of legally enforceable responsibilities. Workers were not the only ones caught in the emerging web of capitalist relations. Directors in distant London beholden to stockholders and their Agent in the Punjab had an agenda different from, and partially independent of, the British officials of the Punjab Commission. Sultan lost much of his manoeuvring room; not even John Lawrence could save him in the long run. The substantial Indian contractor was often a man caught, not between, but in two worlds. Indians were essential to the management of railway construction in India, but when they stood forth like Muhammed Sultan and Jamsetji Dorabji and stood in a direct relationship with a railway company or the government as the main contractors, they did take large risks—more than their European counterparts—which, along with the novelty of some aspects of the construction process and an initial reluctance of the railway companies or their engineers to accept Indian bids, may help to explain their limited numbers in the first phase.

The Punjab material discussed above is fascinating in its richness and complexity. Though only dealing with 251 miles of track, it reveals much about the complex, far-flung assembly of men, material and plant that went into Indian railway construction: plant, bridge girders and rails laboriously shipped north from Karachi; labour mobilized; timber acquired from Kashmir and ballast from, among other places, that most ancient of South Asian urban places, 'the old mounds at Hurruppa. . .'.[90] Moreover, because of idiosyncratic features of the Punjab situation—conflicts between a Chief Engineer and an Agent, the activist presence of Punjab officials, the standing forth of Sultan—we get a longer glance at the Indians in the records.[91] Indian 'managers' were present in construction everywhere in India, but they are usually invisible to historians because they did not make it into the government proceedings.

There still remained for construction the longest link in the SP&DR's jurisdiction: the 303 miles from Amritsar to Delhi which included the

[90] IOL&R, L/PWD/3/62, Bengal RR Letters, no. 79 of 1863, enclosures. The full quotation, from a progress report by the Chief Engineer dated 30 June 1863, is: 'Unfortunately a dispute arose respecting the ownership of the old mounds at Hurruppa, from which we had obtained a considerable supply of ballast. . . .' Railway construction speeded up the deforestation of the subcontinent; it also helped to destroy some of its oldest antiquities.

[91] Chief Engineer Brunton had been fired by Agent Raeburn who in turn was dismissed by the Railway Company at the insistence of the Government of Punjab. Contracts and contracting had figured in the dispute between the two men.

crossing of a number of major rivers. The Government of Punjab still argued against letting the entire construction to a large English contractor. It recognized that overly large contracts to Indians had been let on the Multan line, but on the Delhi line, where labour was going to be more plentiful, government argued that the building should proceed through small contracts to Indians superintended by the engineers of the SP&DR. Mindful of the previous difficulties, the Company chose a different pattern. They called in London for tenders to build the entire line. This resulted in four bids by the summer of 1863: in ascending order, those of Swan and Musgrave; Brassey, Wythes and Henfrey; Peto and Betts; and Waring Brothers and Hunt. The low bid of Swan and Musgrave, who had previously held some small contracts on the Sind line, was dismissed as being 'entirely beyond their experience and resources. . .'.[92] Thus, Brassey and partners, who gave the second lowest bid, got a second crack at building railways in India.

Brassey, Wythes and Henfrey signed a turnkey contract (including rolling-stock) in May 1865.[93] Henfrey, again serving as the resident partner, arrived in India with a large staff of assistants (and, no doubt, with the hard won lessons of the EBR firmly in mind) early in 1865. The stipulated date of completion of May 1870 was met except for a stretch across the Sutlej valley slowed by a failure of the viaduct in the floods of 1869, and the subsequent decision to lengthen the bridge by an additional one-half of a mile. Despite this, a through connection was available in October 1870 after the first half of the line, Delhi (Ghaziabad) to Ambala, had been opened with great ceremony in November 1868 by none other than John Lawrence, by then Viceroy of India.[94] Henfrey et. al. must have learned their lessons well. Henfrey reported that 'throughout the execution of the contract a sufficient amount of labour was at all times obtainable'.[95] He made no remarks about unexpected increases in the cost of labour, and the contract, no doubt with protective estimating, appears to have been a profitable venture for the partners.

Meanwhile, the early phase of Indian railway construction was being conducted according to a different pattern by the MR and the GSIR in

[92] IOL&R, L/PWD/5/1, Public Works Old Series, collection 4A: 'Delhi Railway. Correspondence Regarding the Tenders for the Construction of the Delhi Railway'.

[93] Davidson, pp. 314–24 and Berridge, p. 42 [Berridge has an incorrect date] provides the basic details.

[94] Helps, pp. 275–7.

[95] Quoted in Helps, p. 276. How it was obtained it is a different question. See chapter four of this book.

South India, and the BB&CIR in western India. In the early going these companies, for the main part, chose to avoid the use of large contractors and were in favour of building their lines through the efforts of their own engineers. This system of construction management, known as the 'departmental system' or simply 'departmentally' (and sometimes as the petty contract system, though that was not an entirely accurate designation) basically meant that the engineers functioned as contractors. The MR—and subsequently the smaller GSIR—adopted the departmental system because the first CE, George Barclay Bruce, 'had had in Bengal experience of the annoyance caused by feeble and incompetent contractors. . .'.[96] In the case of the BB&CIR, the main influence appears to have been that of the line's main promoter and consulting engineer, Lieutenant-Colonel J. Pitt Kennedy, who believed strongly that constructing railways departmentally was the most economical method.[97] Kennedy's background as a Royal Engineer in the PWD probably predisposed him to the departmental system just as Berkley's prior experience led him to favour the large contract system for the GIPR. Therefore, the decisions in the start-up decades to adopt one or the other system were in considerable measure a function of timing and of the personal backgrounds of the chief engineers and others with influence. The initial efforts of the GIPR and the EIR were influenced by the British example of large contracts. The failure of that system on the EIR that Bruce had experienced led to the use of the departmental system on the MR. The subsequent failure of later efforts on the MR to introduce substantial contractors reinforced the dominance of the departmental system in the South while the mixed results in the West, North and East were reflected in later decades in those areas where both systems continued in use.

The railway engineers sub-contracted with Indians or Europeans for a specific piece of work—perhaps a very small section of line, a bridge, the supply of sleepers, bricks, lime or ballast, or the completion of so much masonry work, embanking or earth removal.[98] The 'sub-contracts', often little more than verbal agreements, were sometimes not for a specific job

[96] Davidson, p. 344.

[97] Ibid., p. 292. Kennedy was the same man, then a major, who served briefly as the Consulting Engineer to the Government of India for Railways and whose memorandum on a general system of railways for India helped to shape Dalhousie's thinking when the latter wrote his minute on railways in 1853. Kennedy resigned his post in 1852, returned to England and started his extensive involvement with the BB&CIR.

[98] Longridge, p. 3, referred to petty Indian contractors as 'men nearly in the condition of our [British] gangers'.

but for work executed at specified rates. Thus, rather than a contract to do 10,000 cubic feet of earthwork by a certain date and at a certain price, a sub-contractor would undertake work at so much per 1000 cubic feet and would be paid promptly according to measured performance, although advances could be supplied to enable these men to obtain and retain labour. These sub-contractors, in turn, would often make engagements with gangs of earthwork coolies, masons, carpenters, brick-makers or tree-fellers to provide a certain amount of labour or materials.

Separately or in combination with petty contracts, the engineers also employed what was called day labour and the distinction between day labour and petty contracts is maintained throughout the half century covered in this study. Day labour was paid off at the end of each working day—that is clear enough—but what is less clear is by whom, or through whom, the hiring was done. Perhaps the engineer (or more likely his overseers) cut out the sub-contractors and dealt directly with the headpersons who commanded labour. This is certainly the most likely possibility. Sometimes the coolies themselves may have been recruited as individuals and not through some headperson, however petty. This latter possibility is suggested by the sources and, although it must be approached with caution, provides evidence of a railway-led, labour market penetration into some of India's villages in the 1850s. George Bruce wrote in early 1854, in a fashion that presaged Frere's later paean:

> The appearance of Europeans in so wild a District—unarmed with power—vested with no authority—unconnected with Government—was calculated to excite the astonishment and suspicion of the inhabitants; and when it was found that all labour was to be procured irrespective of the heads of villages, by holding out to the Coolies themselves the inducement of a 'fair day's pay for a fair day's work', there instantly arose on the part of all the inferior native officials, an amount of petty persecution, and passive obstruction, which can only be understood by those who have experienced it.
>
> I am happy to say that this opposition has been invariably broken down by the perseverance of the Engineers; and the people along the course of our line begin to understand the blessings of being able to carry their labour to the best market, without seeking the sanction or caring for the pleasure or ministering to the avarice of some petty tyrant. A result, the beneficial effects of which, in elevating the position of the people, we can scarcely calculate[99]

Later, Bruce supports this general statement about day labour by noticing a particularly under-populated area where work by petty contract was

[99] IOL&R, V/23/143. *Selections from the Records of the Madras Government.* no.18: *Report of the Railway Department for 1854* (Madras, 1855), p.3.

impossible and where nearly the whole had been done by daily labour with the tools being provided to the people. 'This involves a large amount of minute supervision, and great personal skill on the part of the Engineers and Inspectors, which I hope will presently be remedied.'[100] Some corroboration for this direct, individual recruitment of day labourers is found in the records of the BB&CIR. In October 1858, the CE of that line, A.W. Forde, authorized the private circulation of a short 'Memoranda For the Information and Guidance of Those of the Engineering Staff Engaged on Construction'.[101] This 18 page item, an example of an early attempt to routinize aspects of the construction process, details the procedures for earthwork, brick-making, brick masonry and other activities. It argues that the engineers should seek petty contractors for the supply of lime, fuel and other materials but that experience had shown that 'the defective arrangements' of the local people made it impossible for them to 'compete in any construction with the Company's Engineers, supported by a well-organized staff of superintendents and labourers'. Forde advocated an elaborate system of supervision designed to give the engineers control of the work: muccadums appointed by the engineers (rather than, as was customary, by the chief Indian overseer) who had direct supervision of the work and who were responsible for checking the tools going in and out every day; mistries of different castes to check one another; overseers to count and record daily the size of gangs plus timekeepers and check-clerks; and finally engineers and assistant engineers who were never to refuse personally to check the task work of the labourers if they felt themselves over-tasked or improperly measured. Does all this not speak of an attempt to more effectively subsume labour under capital? The memoranda does usually speak in terms of gangs doing this and that; for example, the expectation that a day's work would involve a specific task, a certain cubic quantity of earth being moved by a certain number of men, for which they were collectively paid. On the other hand,

[100] Ibid., p. 5. I cannot ignore this evidence but, I must admit, it makes me uneasy. The direct hiring by Europeans of individual villagers to build railways runs contrary to what I think I know about mid-nineteenth-century India and to what most of the other railway sources seem to tell me. In any case, Bruce thought it unusual and hoped the situation would be remedied.

[101] IOL&R, L/PWD/2/121, Railway Home Correspondence—C. Letters to and from Railway Companies. Register—IV—Bombay, Baroda and Central India Railway Company, 1859. A member of Forde's staff, a Mr Jacob, wrote the material. The memorandam was reprinted in *The Bombay Builder*, IV (August–November 1869), pp. 390–3 which suggests its continued use as a manual by the engineering staff of the BB&CIR.

the minute supervision, provision of tools, daily payments and absence of even the pettiest of contractors, were all similar to Bruce's description of some MR construction—what is absent is the praise of the labourer 'freely' and individually recruited to the work-site, and what is present is usually a reference to gangs as the smallest unit, except for brick-makers, who are referred to in a fashion that suggests individuals.

Regardless of the extent to which Indian construction labour was, or became, free or not—and the choice is not antinomian—the departmental system of construction involved more and a different kind of work for British engineers.[102] Their role as technical consultants and supervisors, as professional engineers, was subordinated to their role as managers of the co-ordinated use of large bodies of labourers and to the whole apparatus of accounting and record-keeping that went with it.[103] It was not to many engineers' taste, since 'the Departmental system entails heavy drudgery in the keeping of detailed accounts, and also pecuniary responsibility of a nature foreign to the customs of their profession to which they of course object', but it was always the default position in Indian railway construction: in the 1850s and 60s as illustrated above, and later also, it was the engineers who completed the works of failed contractors.[104]

Thus we find present in the early phase of railway construction the two basic patterns of construction management—construction by substantial contractors, and construction departmentally by the railway companies' (later to include state railways) engineers—which were to persist for the remainder of the period covered in this study. These two basic patterns, in turn, lent themselves to varying combinations and sequences of combinations, with the choice being situationally determined. Moreover, there were differences in what constituted a large contract—certainly some substantial contracts were let which did not involve entire lines. The main dividing line between the substantial contract (however small) and the petty contract (however large) was the principal from

[102] I do not want to get into the morass of the power of constructed discourses, but relevant here is Gyan Prakash, *Bonded Histories: Genealogies of Labour Servitude in Colonial India* (Cambridge: CUP, 1990).

[103] It is the departmental system that, in most instances, left the most detailed record of the labourers and the labour processes for historians to consult. Conversely, an investigation into bridge failures on the Delhi–Amritsar line was hampered by 'the dearth of any recorded chain of incidents' due to 'the works having been executed entirely by Contractors'. IOL&R, P/590, PWD, RR Progs., appendix to February 1875, no. 55.

[104] MSA, PWD (Railway) 1858, vol. 71. Compilation no. 353: 'Railway works. Contract and Departmental System of Executing'.

whom the contract was received: a substantial contract was a first-order contract authorized by a company and a chief engineer or the government and a chief engineer.[105] Substantial contractors often had the autonomy to sub-contract (though later contracts tended to restrict this by requiring the CE's approval of any sub-contractors), but they also had legal binding contracts that made them responsible for the completion of their contract by a certain date. Usually, no written document passed between a petty contractor and an engineer. But once you got down to a certain level in the construction process, the options were similar. Thus, departmental construction might mean the use of petty contractors and/or daily labour. Substantial contractors could and did sub-contract and the agents (at the higher level often professional engineers whose career paths had brought them into the employ of a major contractor) of those contractors might have occasion to resort to daily labour. A failed contractor might be followed by another contractor who succeeded (e.g. in the Bhore Ghat case) or by departmental construction. The one sequence which rarely occurred was for a contractor to take over a project begun departmentally.

Both patterns had their dedicated proponents and opponents.[106] Both sides argued that for economical yet quality construction, their method gave the best results. There appears to have been a disposition to build state railways departmentally and since state construction did not begin until 1869 this helps to explain why the large contracts cluster more heavily in the first two decades.[107] State railway construction through the 1890s was

[105] The Government, of course, through its supervisory and regulatory powers also had a role in the granting of contracts issued by the guaranteed companies.

[106] Berkley, *MPICE*, 19 (1859–60), p. 606, writes glowingly of the beneficial results of the GIPR's use of the contract system. He did not live to see many of the GIPR bridges fail taking with them some of the more positive views of the benefits of the contract system. IOL&R, P/1195, PWD, RR Progs., May 1878, no. 44 has a 'memorandum by T.W. Armstrong . . . Chief Engineer, Central Provinces, upon the major and minor contract systems as applicable to the Central Provinces and to State Railway Works, Nagpur and Chattisgarh Districts in particular', dated 11 May 1876, that provides a reasoned defence of the departmental system and an interesting exploration of the idea that 'the employment of large Contractors in England and in India are not parallel cases'. Armstrong took a shot at the GIPR which, to those aware of the details, was 'a warning and an example to avoid that system in India'. He was undoubtedly referring to the allegations of scamping by the contractors. A generalized attack on PWD inefficiencies and the departmental system by a former member of the Department, Lieutenant-Colonel Tyrrell, can be found in E. Bell and F. Tyrrell, *Public Works and the Public Service in India* (London: Trubner & Co., 1871), pp. 4 and 20–21.

[107] Horace Bell, *Railway Policy in India* (London: Rivington, Percival and Company, 1894), p. 103.

the responsibility of the Public Works Departments (Imperial and Provincial) within which engineers moved among various works of civil engineering and where, in the construction of roads and canals, they would most often use the departmental system. Perhaps, too, for career engineer-bureaucrats there was the wish to control as much as possible, and to maximize the recognition that came from unshared success. There developed, as well, a certain animus among senior PWD officials against large contractors. Indeed, General Strachey asserted in 1884 that 'if there is one thing that has become more and more obvious . . . as regards railway construction in India, it is that construction by contract is being entirely given up as unsatisfactory'.[108] The historical record provides partial support for Strachey's view although large contracts were not given up entirely. Construction through the medium of large contractors was less common in the 1870s and 80s than it had been in the 50s and 60s. Neither the private Bengal–Nagpur line nor the East Coast State Railway, two major lines whose first portions were sanctioned respectively in 1884 and 1890, used the large contract system. The standardized reports on projects for the construction of new lines in the 1890s more and more anticipated the use of petty contractors supervised departmentally. For example, projects as diverse as the 273-mile East Coast section of the Bezwada–Madras Railway, the deep south Pambam Branch of the SIR, a proposed Bombay to Sind connection, and a Ranaghat–Ganges–Katihar line were all said to be best carried out by readily available petty contractors.[109]

The increased availability of competent petty contractors reveals a major consequence of railway construction and an important, evolutionary development in the construction process. Petty contractors had become a numerous presence within Indian society. A class fraction of petty capitalists had emerged out of the earlier opportunities presented by railway and other construction work.[110] Striking testimony of this

[108] *PP* (Commons), 1884, *Railway Communication*, para. 71. He meant large contracts. During his long career General Richard Strachey held many of the senior positions in the Government of India, including Secretary, PWD. He became chairman of the EIR Company in 1889.

[109] IOL&R, P/4586, PWD, RR Construction Progs., May 1894, no. 406; P/5914, February 1900, no. 163; P/5915, March 1900, no. 82; and P/5918, July 1900, no. 217. These standardized project reports are another example of the routinization of the construction process.

[110] For a brief discussion of this process among a north Indian caste group, the Nuniyas, see William L. Rowe, 'The New Cauhans: A Caste Mobility Movement in North India', in James Silverberg (ed.), *Social Mobility in the Caste System* (The Hague: Mouton, 1968). esp. p. 70.

ɔpment was provided by an engineer engaged in doubling part of the ı Western Railway (hereafter NWR) in the late 1890s who compared his situation with that encountered by Brunton in Sind some forty years earlier.

In the place of the coaxing Mr. Brunton and his colleagues had to bring to bear on the natives to induce them to take up work, we now find ready to hand a recently created class of middlemen or petty contractors thoroughly conversant with the actual cost of work, and in fierce competition. The engineer's difficulty is now not so much to find contractors as to weed out the bad hats and men of straw who tender at impossible rates, whilst fully aware that they must fail to carry out the work honestly.[111]

Most of these men found their way in the emerging relations of capitalism as those who could provide and manage labour power. Their increased presence helps to explain why the supply of labour ceased to be an issue in most railway construction. The petty contractors became the activators and linkers within the increasingly wide-spread labour market(s). The construction process was facilitated to the extent that these petty contractors became conversant with the work and disciplined to the demands of the market. Experienced contractors were as important as trained workers. Many of the failures in the early decades of railway building were the failures of management: inadequate contractors at many levels and lax supervision by engineers.

Apart from the reasons mentioned above, the transition may also have been related to the passing of the era of the great railway contractors in Britain, while in India, as Major-General Dickens testified to a House of Commons Select Committee in 1878, the contract system was 'of modern introduction'.[112] They sometimes had good, reliable contractors to build railways but the experience had been mixed; it would take some time for the contract system to become general and for competition among reliable firms to become widespread, so in the meantime the PWD needed the capacity to build departmentally. Thus, in the 1870s and early 80s, the IVSR, the Punjab Northern State Railway (hereafter PNSR), the Oudh and Rohilkhand, and part of the Rajputana line, among other State ventures, primarily were built departmentally—but so was much of the private Indian Midland Railway.

[111] ICE MSS. no. 3069, Victor Edgar De Birchin De Broe, 'The Doubling of the North-Western Railway of India', 1897.

[112] *PP* (Commons), 1878, cmnd. 333, *Report from the Select Committee on East India (Public Works)*, paras. pp 517–20.

1. Left training bund, Curzon bridge site over the Ganges river at Allahabad, *c.* 1903.

2. Pathan workmen, *c.* 1880.

3. Bell's patent dredgers, Sher Shah bridge over the Chenab river , *c.* 1889.

4. Loading a well to force it to sink, Curzon bridge, *c*. 1903 .

5. Girder erecting, Landsdowne bridge over the Indus river at Sukkur, *c.* 1888.

6. Lime-kilns, Bezwada bridge, *c.* 1891.

7. Mortar mills, Sher Shah bridge, *c.* 1889.

8. Platelaying in the Peshawar City station, mid-1880s.

9. View of the Bhore Ghat and the reversing station, early 1860s.

10. Masonry bridge, Bhore Ghat, *c.* 1860 .

However, there was railway building in India by large contractors throughout most of the period covered in this study. Sometimes, what determined the choice of method was not policy or individual preference but local exigencies, such as the backgrounds of available personnel. Thus, the Gwalior–Jhansi section of the Indian Midland Railway, a difficult stretch, was contracted out despite the Chief Engineer's inclination to do the work departmentally. The reason was that the staff of engineers available to him had 'little or no knowledge of the language of departmental methods of works and accounts' . . . and he did not feel that he could '. . . safely entrust the prosecution of the work to them except through the agency of a Contractor'.[113] Ironically, the contract went to Glover and Company. Thomas Glover had testified before the same Select Committee which had heard Strachey pronounce that the contract system was being abandoned.

Thomas Craigie Glover, a Scottish engineer turned contractor, in conjunction with various partners, was the most enduring of the large contractors from Britain to maintain a presence in Indian railway building from the mid-1860s onwards. By 1884 Glover had made 42 extensive visits to India in order to obtain and execute contracts.[114] He did extensive works for the Bombay municipality and became involved with the GIPR as the builder/rebuilder of bridges, including the Mhow-ke-Mullee viaduct. He also built 40 miles of the BB&CIR (illustrating again the mixture of patterns since much of the BB&CIR was built departmentally) and took the contract to double the 70 mile stretch of the GIPR between Igatpuri and Bhusawal.[115] He then took extensive contracts to build State Railways: 250 miles of the Rajputana Line from Agra to Ajmer, 75 miles of the Sindia State Railway, Agra to Gwalior, the heavy stretch of the Bhopal State Railway from Itarsi to Bhopal, and the 60 miles of the Indian Midland, from Gwalior to Jhansi. He always completed his contracts on time (there was a slight delay on the Rajputana line which was not his fault—a famine in Bengal required the diversion of material for the permanent way), though not without acrimonious disagreement with some PWD chief engineers. The disagreement reached a peak in the Bhopal contract and resulted in an arbitration and in Glover's cutting testimony to a House of Commons Select Committee, where he lam-

[113] IOL&R, P/2750, PWD, RR Construction Progs., July 1886, no. 151.

[114] *PP* (Commons), 1884, *Railway Communication*, paras 1827–37.

[115] Work of reconstruction and line doubling which would have employed significant amounts of labour.

[...]ne PWD engineers whom he characterized as having limited practical experience in railway work and who were tied up in red tape and prone 'to hold up the specification as their Bible. . .'.[116]

The fact that eleven firms bid for the Bhopal contract and eight for the Indian Midland contract suggests that Dickens' expectation that more competition among contracting firms would begin to appear, proved correct.[117] The government's engineers were able to choose on the basis of price comparisons and, in some cases, on the basis of the past performance of the firms. In both cases, the latter worked to Glover's advantage, since in neither case was his bid the lowest. In both competitions, a number of Indian firms were present. Contracting in various spheres, we know, was to become the basis for a good number of Indian fortunes in the later nineteenth and twentieth centuries.[118] Although Indians did not win either of these contracts, a number of Indian firms did win substantial contracts on other lines.

Other contractors took up major works. Brassey, Wythes and Perry built 147 miles of the EIR chord line.[119] The conduct of this contract, the third in India with which Brassey was associated, came in for serious criticism. Perry, the resident partner (a former EIR engineer), was accused of running his 'operations in a spirit of meanest parsimony, which, unless speedily reformed, must result in great discredit to Messrs. Brassey and Company'.[120] This statement reinforces my view, expressed earlier, that some of the great British contractors who partnered to hold a railway contract in India had a limited, direct connection with the execution of the contract. When Henfrey, who had a certain reputation of his own, managed Brassey and Company's Indian contracts, the resources of the partners were extensively committed. When Perry, a lesser light, was in charge, the situation changed.

[116] *PP* (Commons), 1884, *Railway Communication*, paras 2061–4.

[117] IOL&R, P/1524, PWD, RR Construction Progs., December 1880, nos. 77–102; PWD, RR Construction Progs., July 1886, nos. 148–51.

[118] Commissariat contractors for example. See Clive Dewey, 'Some Consequences of Military Expenditure in British India. The Case of the Upper Sind Sagar Doab 1849–1947', in C. Dewey (ed.), *Arrested Development in India* (Riverdale: The Riverdale Company, 1988), p. 157.

[119] Helps, p. 166; but G. Huddleston, *History of the East Indian Railway* (Calcutta, 1906), p. 42 remarks: 'there were great difficulties with the contractors who had taken up the work of construction and it was on this account that the chief delay occurred.' At the same time Brassey and partners took the contract to build waterworks for Calcutta.

[120] IOL&R, L/PWD/3/66, Bengal RR Letters, no. 39 dated 4 April 1867, enclosure. The letter and the enclosures all testify to the poor progress of the works.

Another major partnership, Waring Brothers & Hunt, expeditiously completed the Jubbulpore extension of the EIR, under the tightly-drawn specifications of the chief engineer, Henry Peveril LeMesurier, and the watchful, well-organized supervision of his subordinate engineers. Le Mesurier (1828–89) was born in Guernsey, the eldest son of Benjamin Le Mesurier, whose second son, Charles (1829–77), also served as a railway engineer in India.[121] Henry Le Mesurier made a name for himself in India and subsequently went on to a distinguished career elsewhere. He articled for five years with R. B. Grantham, and then served the contractors, Jackson and Bean, as an engineering assistant for breakwaters and forts which they were building for the British Government on Guernsey. In 1853, he accepted an appointment on the EIR as an assistant engineer and reached India in early 1854. He worked on surveys and the construction of the EIR main line, performed valuable service during the mutiny and rose rapidly. In 1859, at the age of 30, he was appointed CE for the Jubbulpore extension, the survey parties of which he had been organizing. Ill-health drove him to England subsequently, where he remained until July 1863 as he could effectively prepare specifications, etc. in London. He remained CE of the Jubbulpore line until 1868 and, after a few months as CE for the PNSR surveys, joined the GIPR as CE of its Presidency Division. By October 1869 he was the agent of the GIPR, a position he held until February 1877, when he left India to become a member, and later President, of the Board of Administration of Egyptian Railways. Health problems led him to return to Guernsey in 1887 for recuperation, but while there he had an attack of paralysis which made his return to Egypt impossible. A second attack took his life in July 1889. Le Mesurier was representative of a group of British railway engineers who went to India early in their careers, made a name for themselves and then went on to achieve greater distinction elsewhere.

Charles Innes Spencer, Le Mesurier's second-in-command, and the acting CE in his absence, was also a rising engineer. He lived his childhood in France where he showed great aptitude for mastering foreign languages—five by the age of eighteen.[122] Later, in England, he demonstrated a strong mathematical ability and precociously pursued studentship in engineering at King's College, London. He sailed to India in 1850 to find employment, which soon came his way. Successively, more important postings led to his appointment under Le Mesurier. Spencer had

[121] Obituaries of Charles and Henry can be found in *MPICE*, 51 (1878), pp. 266–8 and 98 (1889), pp. 393–7. Their grandfather Henry was the hereditary Governor of Alderney on Guernsey

[122] Obituary, 'Charles Innes Spencer', *MPICE*, 43 (1875–6), pp. 304–6.

engineering ability, resourcefulness and inventiveness. An example of this is when he found the Jubbulpore line workmen unwilling or unable to knock away the supports of arches simultaneously, 'he adopted the plan of resting the ends of the timbers on strong bags of canvas, filled with sand, to be emptied by turning a tap, regularity in this light action being secured by giving time with a flute'.[123] After the line was completed he was appointed its district engineer, in which position he proved to be able and well-liked. An accident in the summer of 1875 contributed to his premature death in November of the same year.

The Waring Brothers were the sons of John Waring, a well-known contractor for public works in England.[124] They carried out major railway contracts throughout the world. It is unlikely that any of the Warings visited India during the life of the contract. They provided the reputation and the financing, while Hunt provided the connection with India. James Hunt was one of the few who successfully completed EIR contracts for the Experimental Line, and subsequently in the Mirzapur district.

The contractors' chief agent was Mark W. Carr. Carr (1822–88) had served an apprenticeship in Robert Stephenson's firm and had then gone on to hold important construction and open-line positions in railways in Britain. In 1858, he was appointed CE of the GSIR, a position he held until he took up the Jubbulpore line construction.[125] After the Jubbulpore line he built a railway in Hungary and, from 1873 to 1879, served as the General Manager of the Rio Tinto Mining Company in Spain, following which he became Managing Director of the Blaina Ironworks while continuing to travel widely in his role as a consulting engineer. He died in Mexico where he was inspecting some silver mines on behalf of interests in England. Carr was assisted by another senior agent, George Nicoll, a former EIR District Engineer who had assisted the Hunt and Elmsley partnership to complete their Mirzapur contract. Nicoll then served on the Jubbulpore survey and acted as the project's CE for a brief period after Le Mesurier's health drove him to seek relief in England. The Railway Company permitted Nicoll to resign in December 1862 in order to enter the employ of the contractors.[126] Richard Shaw Brundell (1829–1903) and John M. Easton, two other trained engineers, acted as the contractors' agents in charge of the Jubbulpore district. Brundell, too, had worked for the EIR previously. After the Jubbulpore contract, Brundell and Easton

[123] Ibid., p. 305.

[124] Obituary, 'Charles Waring', *MPICE,* 92 (1888), p. 410. He died in 1887.

[125] Obituary, 'Mark William Carr', *MPICE,* 93 (1888), pp. 487–8.

[126] IOL&R, P/217/37, NWP, PWD, RR Branch Progs., 1 March 1864, 'Contract for the Construction of the Jubbulpore Line'.

formed a contracting firm of their own, based in Jubbulpore. Their first contract was to maintain the Jubbulpore line for two years and to complete extensions and additions to the stations. They subsequently obtained a similar contract for 120 miles of the GIPR south-westward from Jubbulpore and a large contract to supply teak sleepers, the raw material for which came from the Rewah forests.[127] The men discussed in this paragraph demonstrate the presence of interconnected opportunities for engineers that could include employment as a railway company's engineer, a contractor's agent, or an independent contractor.

Another set of agents, Hamilton, Brown and Company of Mirzapur, were responsible for receiving the permanent way materials at Mirzapur and forwarding them to the line via the Great Deccan Road. The functions of these agents point to another important spin-off of the construction process, namely the cartage of plant and materials to construction sites. There were numerous political, legal and commercial considerations and activities preceding actual building. For instance, since the Deccan Road ran through the state of Rewah, negotiations had to be conducted, through the resident political agent, with the Rewah Chief to ensure the smooth passage of railway materials and their exemption from transit duties levied by the small, princely state.[129]

These biographical details illustrate a number of points. First, experience of Indian conditions and the quality of the senior people among the Company's engineering staff and the Contractor's staff, help to explain the speedy success of the Jubbulpore project. Also, their shared status as civil engineers reduced friction between them. Second, the details illustrate the routinization of the construction process. A pool of men experienced in railway engineering in India had come into existence and they, moving sequentially from project to project and from railway companies to contractors, provided the engineering and organizational expertise which was often lacking in the 1850s. The construction process, viewed as an assembly of sub-processes, became more organized, and contractors at various levels subject to more effective supervision of engineers.

In summary, the management of Indian railway construction followed

[127] Obituary, 'Richard Shaw Brundell', *MPICE*, 154 (1903), pp. 386–7.

[128] IOL&R, P/217/37, NWP, PWD, RR Branch Progs., 25 February 1864, 'Transit Duty on Permanent Way Materials'.

[129] When a major princely state was involved, the negotiations could be very complicated indeed. See Bharati Ray, 'The Genesis of Railway Development in Hyderbad State: a Case Study in Nineteenth-Century British Imperialism, *IESHR*', 21:1 (January–March 1984), pp. 45–69.

two patterns: the large contract system and the departmental system. The large-contract system was responsible for a substantial amount of the miles built between 1850 and 1900, although the greater part of that work was done in the period from 1850 through the mid-1870s. This system came in a variety of forms which, though they did not alter the essential pattern, did have an impact on performance. One variation was between those who took contracts as an exercise in financial speculation, and those, whether new to contracting or of established reputation, who had a commitment to seeing their contract through to completion. A more subtle difference was the one between partnerships in which a resident partner was present on the contract, and one who left the execution of the work in the hands of a managing agent. The parade of GIPR engineers, some very senior, who left the Company's service to become contractors' agents, could be an example of the latter; it could even be an example of the former, since we do not know for sure whether those senior engineers turned agents were salaried employees or not. If they were commission agents receiving a percentage of the profits, they could have been motivated to cut corners, or if sub-contractors, to cut costs.[130] Another difference which cross-cut the two just mentioned, was between British-based contractors, among whom were some of the most prominent Victorian railway contractors, such as Brassey, Peto, Wythes, the Waring Brothers and those who were based in India. The latter group was further divided into those composed of Indians and those composed of Europeans resident in India. Except for Indian labour, the great partnerships could assemble more easily 'the factors of production', including, in the case of turnkey contracts, the provision from Britain of iron rails and rolling-stock (Brassey, for example, owned iron works).

The departmental system was more uniform. Considerations of profit and financial speculation did not affect the performance of the engineers, though they were often under pressure from a railway company or the State to build economically. Important determinants of performance under the departmental system were the experiences and abilities of the engineers involved.

A major difference between the large-contract and the departmental systems was to be found in the added layer of supervisors in the former. The contractor executed his contract through agents who were, for the main part, engineers; they, in turn, had their staff of European and Indian

[130] C.B. Ker and R.W. Graham, who had served as CEs of the GIPR, became contractors' agents.

overseers. These people functioned in the same fashion as their counterparts in the departmental system. In the large-contract system, the engineers and overseers of the railway company or the State railway system supervized and checked the work performed for contractors; they made sure that the work was done according to design and specifications and they authorized payments to contractors based on properly completed work. This dual system that freed the engineers to concentrate on technical supervision was considered a benefit by the advocates of the contract system.[131] Opponents of the system claimed it added an unnecessary intermediate layer since, for the main part, the contractors' agents—like engineers under the departmental system—had to hire Indian subcontractors. Both patterns, it must be stressed, included that upper layer of supervision represented by government consulting engineers and those within the State railway system who functioned in the same capacity. In one way or another the Government of India or its subsidiary jurisdictions (and often both) maintained a close watch over the construction process.

The conduct of management is an important factor in the evolution of capitalism. Management attempts to control the nature and conduct of work, and higher management seeks more effective control of lower management. A good part of the routinization of the railway construction process involved precisely that: better control of contractors and subcontractors by engineers and government officials (PWD engineers being, in effect, both). This chapter described some of the many ways in which more effective co-ordination of the overall process and better control at or near the point of production were achieved. Particularly instructive at the level of co-ordination was Le Mesurier's supervision of the Jubbulpore extension, while the appearance of more and more competent petty contractors, as in the NWR reconstruction in the 1890s, illustrated the process of better control at the point of production.

The tasks of management were complex and far flung. The inter-continental and sub-continental assembly of workers, material, and equipment was sorted out within the complex hierarchy of supervision, direction and work-site management sketched in this and the preceding chapter. In terms of the mobilization and utilization of labour, management came to inhere primarily in various levels of contractors and sub-contractors, regardless of whether the departmental system was used or not. Relations in production largely involved the interactions of Indian workers

[131] 'The Contract System', *The Bombay Builder*, vol. III (5 September 1867), pp. 93–4.

with other Indians who directly commanded their labour. Though by nineteenth-century standards the British broadly conceived, extensively co-ordinated and closely supervized the overall process of railway construction in India, they managed only indirectly and largely through the authority of Indians at the point of production. The British had little choice.

Management within capitalism takes many forms.[132] Sub-contracting was well-suited to the conditions of railway construction in nineteenth-century India, as it continues to be in the construction industry—and a good many other industries—in the developed capitalist countries of the late twentieth century. Moreover, in India, in addition to all the standard divisions between capital and labour, higher management and labour were separated by cultural chasms. Sub-contracting helped to overcome that problem since below a certain level Indians managed Indians. Contracting and sub-contracting, however, did not hinder the shift towards a more effective management of the overall railway construction process. The presence of more and more experienced petty contractors vastly reduced the problems and delays at one level, while at another level, as part of the same development, British engineers learned to engineer men better and understand better the technical challenges India's terrain and climate presented to the railway engineer.

[132] Usefully surveyed in Sidney Pollard, *The Genesis of Modern Management* (London, 1965).

CHAPTER 4

Obtaining Labour

In getting the railways of India built the British used the labour power of Indian men, women and children. Substantial numbers of Indians became involved in railway construction in particular localities and on an all-India basis: 115,635 were claimed to be at work daily in Bengal in December, 1858 (Table 6), while the aggregate yearly pan-Indian employment estimates presented in Table 2 ranged from a low of 64,134/78,895 in 1867 to a high of 373,212/459,110 in 1898. Furthermore, the direct employment generated by railway construction was supplemented by two other large bodies of workers: those employed to provide construction materials and move those materials to work-sites; and, within a few years of the initial construction, those employed in the reconstruction or large-scale repair of the existing railways. Obviously the general question of how this large workforce was obtained is critical to an understanding of the railway construction process. What kind of labour was needed? Who were the millions of Indians who, over some five decades, physically built the railways? Where did they come from? What were the problems encountered in obtaining such large numbers of workers?

The search for answers to these questions takes us, in the first instance, to the workforces at the construction sites numerically described in Tables 4 to 15. What the tables suggest is that the construction workforce was far from homogeneous; it was divided into many different categories and groupings: age and gender distinctions, unskilled and skilled labour with the latter in turn being divided into different artisan trades, such as carpenters, masons, smiths and so on. The tables do not reveal many other divisions: for example, those based on pay scales, caste, religion, language and region of origin. These various distinctions were important to the recruiting process, to the social composition of the workforces and to the conduct of the construction work. The construction workforce, in short, was fragmented along many dimensions. Recruitment, therefore, must have been a multi-dimensional activity.

An investigation of the age and gender distinctions highlights three important points. First, many of the workers were women and children;

indeed, they were the majority at the Madras sites described in Table 4. Second, men, women and children came to the construction sites as family units. Third, the tables suggest the presence of a division of labour within the families between the men on the one hand and the women and children on the other.

An analysis of Table 4 indicates that of the 5971 people employed on MR district 3 in 1855, 49 per cent were women and 14 per cent were children (boys). In all categories of work, except the laying of the permanent way, women outnumbered men. A later return from district 16 (in the Malabar interior) for April 1857 gives the average employment as 795 men, 7139 boys and 202 women.[1] In 1857, in neighbouring district 15, the deputy consulting engineer, from whose inspection report this information is drawn, stated: 'There appears to be no difficulty in getting coolies to any number required; and the Moplahs hereabouts not only flock to the work (all earthwork as I have said) but they bring their wives and children.'[2] Unfortunately, the Madras authorities, in the period 1855–7, seem to have been the only ones interested in recording the numbers of women and children employed. However, it is clear that women and children were present in construction workforces in many parts of the subcontinent throughout the period under study. Davidson, writing of construction in 1855–8 in the South Birbhum division of the EIR, states 'all the earth is brought by men, women, or children, in small baskets on their heads, from excavations alongside the line'.[3] He goes on to note that, though a seemingly primitive and slow method, it was the custom of the country which proved speedy enough when sufficient labour could be collected. 'It allows, too, the whole strength of a family to be employed, from the grandsire down to the girl and boy of ten or twelve.'[4] Another engineer near Allahabad noted in 1857 that 'women in this country do all the earth carrying . . . and fetching bricks, mortars

[1] IOL&R, V/23/156. Madras Govt., selections from the records, no. 53A: *Report of the Railway Department for 1857* (Madras, 1858), p. 245.

[2] Ibid., p. 245. For more on women in MR construction see IOL&R, MSS Eur. C.378: Typescript copy of 'A Pioneer of the Madras Railway 1867–1875', based on the diaries of Thomas Hardinge Going (1827–75), engineer North Western Line, Madras Railway Company 1857–75, edited by his grandson, Lt.-Col. R.J. Going. The diaries describe survey and construction work between Madras, Raichur, and Hyderabad.

[3] Edward Davidson, *The Railways of India* (London: E & F.N Spon, 1868), p. 162.

[4] Ibid. Davidson says this in the context of south Birbhum but he may have been referring generically to much of the EIR construction about which he had close knowledge.

etc., all is done by women' while the men only did heavy work like laying rails or timbers.[5] Berkley wrote of construction in western India in the later 1850s and early 60s in a similar vein: men, women and children came to railway construction as family work units.[6]

Later evidence shows something similar. An extensive survey of wage rates paid to unskilled, cooly-type labour employed on government PWD projects, including railways, throughout India in April 1878, presented the rates according to the categories of men, women and children.[7] Engineers continued to note women and children labouring at railway construction. E. Monson George's 1894 treatise on Indian railway construction and operation, for example, refers to the earthwork, usually conducted in the dry season, performed 'in a most careless manner' 'by large numbers of men, women, and children, each carrying a basket full of earth on their heads'.[8] The CE of the SIR, in 1905, proposed an elaborate ticketing system in order to pay large numbers of daily-mustered coolies. His categories, each with its own wage rates, were men, grown boys, women, boys and girls.[9] Clearly, women and children were a significant and continuing component of the workforce. They increased the ability of a family to earn a living wage, and also, being lower paid than adult males, helped the railways to reduce their wage bills or the use of more capital-intensive techniques. Indeed, one discovers in this large pool of female and child labour a reason for the large size of the Indian railway construction workforce and for the persistence in India of labour-intensive methods of construction. There was little stimulus for employers to utilize more capital-intensive construction techniques because the cost of labour was low. If one labourer could not earn enough to support his family, then the entire family had to work to ensure their survival. Family labour insured the maintenance and reproduction of labour

[5] IOL&R, Eur. MSS. B. 212, Carrington Papers: Samuel Carrington to his mother dated 15 November 1857.

[6] Berkley, *MPICE*, 19 (1859–60), p. 605. The BB&CIR also employed women and children. See IOL&R, L/PWD/2/121. Railway Home Correspondence—C. Letters to and from Railway Companies. Register IV—Bombay, Baroda and Central India Railway Company. 'Memoranda for the Information and Guidance of Those of the Engineering Staff Engaged on construction.'

[7] IOL&R, P/1197, PWD, RR Progs., October 1878, no. 128.

[8] E. Monson George, *Railways in India* (London: Effingham Wilson, 1894), p.50. On p. 88 he provides a table of wage rates for different categories of workers including women and child labourers who received the same rates, Rs 3 to Rs 4 per month, while the men got Rs 4 to Rs 8 per month.

[9] TNSA, Madras, PWD, RR, 15 April 1905, no. 784.

power when the single wage earner could not. The costs of that maintenance and reproduction had to be borne by the family, since the contractors took virtually no responsibility for the workers' well-being. British engineers, for their part, were slow to take responsibility for the well-being of workers, and slow to recognize that decent housing, good water supply and adequate sanitation could expedite the work of construction.[10]

The unskilled–skilled distinction represented another major division among the construction workers. It was not a clear-cut distinction, but most of the workers fell into the unskilled category. Building a line of railway in India required large numbers of people to move earth and rock. It was direct, hard, manual labour carried on with the simplest of tools and basic contrivances, frequently a basket on a woman's head, for the transportation of earth and rock to or from (embank or cut) the line of way. Once the road-bed had been prepared, ballast also had to brought to the line and spread before the permanent way could be laid—again a task for unskilled labour. Tables 4 through 9 illustrate the preponderance of unskilled labour: 59 per cent of the workers in Table 4 are coolies, women and boys engaged in earthwork and ballasting, and if one makes the reasonable assumption that the women and boys in the bricklaying and masonry category were also unskilled helpers, the unskilled go up to 86 per cent of the total. In Table 7 the excavators and labourers total 103,008, or 87 per cent of the total. Table 9 produces an unskilled majority of 83 per cent. Data from one mile of track from start to finish in the flat conditions of the central Punjab generates similar results—at least 83 per cent of the workers were in the unskilled category.[11] It cannot be an overestimate to say that, overall, at least 80 per cent of the construction workers were engaged in unskilled labour. And, to repeat what was established above, many of these were women and children.

However, there were a number of distinctions within the unskilled category that had implications for the recruitment of labour and the

[10] The presence of families and family work at the construction sites distinguishes many among the construction proletariat from other forms of migratory work in India where the men migrated and the women and children remained in the villages, thus shifting the costs of the reproduction of labour power sold to capital to the agrarian sector. See Gail Omvedt, 'Migration in Colonial India: The Articulation of Feudalism and Capitalism by the Colonial State', *JPS*, 7:2 (January 1980), pp. 185–212. Railway builders directly off-loaded many of the costs onto the working families.

[11] Kerr, *IESHR*, pp. 325–7.

labour process. Firstly, a distinction was frequently maintained between men's work and that of women and children. At one level this could be viewed simply as a distinction that flowed from differential physical demands—men reserved themselves for jobs that required more physical strength. Men's work, however, was better paid, slightly more-skilled and less monotonously simple; it was work, however small the increment, of higher status. If we extrapolate backwards from current studies—in this case with considerable confidence, since there is no basis for thinking that the situation was much different in the nineteenth century—we can infer the presence of hierarchically-ordered gender relationships within these family work units in which women worked at the lower-paid construction jobs and took care of the domestic chores as well.[12]

Secondly, among the unskilled, there was a distinction between the excavators and the labourers. Excavators were those who dug out the earth and rock; labourers, for the main part, were carriers of materials: earth and rock to an embankment, spoil from a cutting, bricks and mortar to work-sites, building materials up scaffoldings and so on. There was always a need for a large number of labourers as fetchers and carriers at any work-site—even the advanced stages of bridge construction needed 'bamboo coolies, or carriers of heavy weights' who lifted their loads up the flimsy staging to the waiting skilled workmen and their helpers.[13] Excavators were usually men; labourers were often women and children.

Thirdly, there was the distinction between those who were particularly effective at earthwork and those who were less effective, whether they worked in family units or not. The former had the skill—and perhaps physique—to accomplish more earthwork per day than others. These were often people, groups of people given the corporate nature of Indian society, whose occupation before the railway age had been earthwork-related: tank-digging or road-building, for example. They were part of that already-established, 'floating' group in the country-side—servants, artisans and labourers—who were mobile and without land.[14] It must be

[12] Cf. Chitra Ghosh, 'Construction Workers', in Joyce Lebra et al. (eds), *Women and Work in India. Continuity and Change* (New Delhi: Promilla & Co., 1984), pp. 201–11. The same pattern—women working during the day and then doing domestic chores in the evening—is described in Jan Breman, 'Seasonal Migration and Co-operative Capitalism. The Crushing of Cane and of Labour by the Sugar Factories of Bardoli, South Gujarat—Part 1', *JPS*, 6:1 (October 1978), pp. 63–6.

[13] Berkley, *MPICE*, 19 (1859–60), p. 608.

[14] Jan Breman, *Labour Migration and Rural Transformation in Colonial Asia* (Amsterdam: Free University Press, 1990), p. 9.

emphasized that especially where earthwork was concerned, the novelty of railway construction in India lay primarily in the substantial increase in the demand for labour whose use had to be co-ordinated and directed on a grand scale. Contractors and engineers valued the contribution of 'trained' earthworkers. Such groups were the shock troops of construction; they led the way; they set high standards of work output. These Indians, though not as brawny or as ferociously effective in cutting and embanking, played a role in Indian railway construction similar to that of the British navvies who, of course, were assisted by better tools, such as bigger spades and large wheelbarrows and were also more roisterous when not at work! These Indian workers provided mobile labour that tramped from one project to another and who, when available, were the leading earthworkers at construction sites. As early as in 1863 the Governor-General, Lord Elgin, observed: 'Bodies of navvies are becoming attached to the companies, who follow them from place to place, and render them comparatively independent of the local supply of labour . . .'.[15] By 1884 the contractor, Thomas Glover, could assert that local labour 'is of little use to us in India'.[16] 'We have always imported the backbone of our labour for earthwork and ballast wherever we are.'[17] However, unlike the single males of Britain, the 'Indian navvy' rarely tramped alone or with one or two mates.[18] He usually came to the work-site in a group composed of fellow members of a caste or tribe (or castes and tribes of comparable status), recruited from the same locality; he also frequently came as part of a family group of which all able-bodied members toiled at the work-sites.

These distinctions had significant consequences for the recruitment of workers. I was long sceptical of Elgin's observation. I was convinced that the majority of unskilled construction railway labour—by far the largest

[15] Elgin to Sir Charles Trevelyan, dated 23/02/1863 reprinted in Theodore Walrond *Letters and Journals of James, Eighth Earl of Elgin* (London: John Murray, 1872), p. 446.

[16] *PP (Commons), 1884*, Cmnd. 284, 1884, *Report from the Select Committee on East India Railway Communication*, para. 1915.

[17] Ibid., para. 2152.

[18] Bodies of construction labourers in India, Canada, or the United States did have other solidarity bonds, such as a shared ethnic background. Navvies in Canadian canal construction, for example, were often French Canadians or Irish. At some Canadian sites, too, the presence of entire families was noted, though the males did the construction work. See William N.T. Wylie, 'Poverty, Distress, and Disease: Labour and the Construction of the Rideau Canal, 1826–32', *Labour/Le Travailleur*, 11 (Spring 1983), pp. 7–29.

proportion—came, in most instances, from the rural populations, particularly the landless agricultural labourers, marginal peasants and poor artisans, living close to the line-of-work; of course, there were areas of sparse settlement with little or no population to draw labour from. I now know that the situation was much more complicated.

Before I examine the specifics of my reconstruction, I believe, in outline, that something like the following occurred: throughout the period under study, unskilled labour was recruited from wherever it could be most easily obtained and reliably retained—neighbouring villages, the wider locality, the region and eventually other regions. However, for reasons examined below, certain kinds of rural labour proved difficult to retain for extended periods. These 'unreliable' workers were those who possessed significant links with the village economy and the village power structure from which they could only be temporarily detached by the lure of construction work, and/or the permission or coercion of village power-holders—coercion occurring when village power-holders undertook to provide labour and became, in effect, very petty contractors. These workers were, in many cases, those obtained from villages near the work-sites since those who came from greater distances had, in that act alone, displayed greater detachment from their village although that detachment was not necessarily sustained.

The more reliable workers were those who had taken up more completely the life of circulating wage labour and who usually came to a work-site from a distance. The ratio of one kind of labour to another varied from work-site to work-site, depending on a changing concatenation of factors which produced changing mixtures of supply and demand. Sparsely populated areas provide limited or no supplies of labour; some projects demanded so much labour that the vicinity could not supply such numbers; poor harvests created more available local labour, bumper harvests had the opposite effect; other types of construction projects competed with the railways for labour, and so on. It is difficult, indeed in most cases impossible, to specify with any precision the ratio present at a given work-site. However, I also believe that from quite early on—certainly by the time Elgin made his observation in 1863 about gangs of navvies—the ratios increasingly shifted in favour of circulating labour. The continued, large demand for railway construction workers stimulated the growth of regional and then inter-regional labour markets within which circulating labour increasingly met the needs of railway companies. This was particularly true for the supply of skilled labourers and unskilled excavators to whom other labourers were attached as family members. Additional labourers and excavators who could not be supplied from the wider pools of circulating labour continued to be sought among the village populations near the right of way.

The railways, along with other public works projects, drew upon the pre-existing, floating rural proletariat and drew those people into widening labour markets.[19] Railway construction also helped to enlarge the number of waged labourers by detaching some from village attachments to join the ranks of circulating labour. This accumulating, decades-long process resulted by century-end in an enlarged rural proletariat existing by construction labour and composed of groups of people who were either in construction-type work prior to the railway age, tribals, or had detached themselves from village life to seek an existence as wage labourers. The climate dictated that much of railway construction in India was seasonal, therefore many migrant workers returned 'home' in the off season; construction work, however, was their main source of livelihood. The pattern of movement of these seasonal workers, therefore, is best characterized as circulation rather than migration.[20] Other workers, however, continued to be found in villages close to the line-of-works and these people were only temporarily inducted into the construction labour market.

Let us take a closer look at the sources of labour, beginning with those the British labelled 'unreliable', and who, however less crucial they became, remained among those who built the railways. Landless labourers, village artisans and marginal peasants were attached to a village in ways English contractors and engineers dimly understood, though comments like the following indicate the beginnings of understanding: 'I think, however, the Zemindars and headmen would come forward as Contractors; in which case there would be no want of labour, as they would be interested in procuring it.'[21]

Many a coolie from this group was not and did not become free in the way that Frere, Bruce and others anticipated; they were part-time proletarians whose main bondage, activated most visibly at planting and

[19] The presence of such people in India prior to British rule is now well established. See Ranajit Das Gupta, 'Factory Labour in Eastern India: Sources of Supply, 1855–1946. Some Preliminary Findings', *IESHR*, XIII:3 (July–September 1976), p. 277.

[20] Breman, *Labour Migration and Rural Transformation*, pp. 48–59. The socio-economic world of these circulating workers moving in family groups was different from the men who migrated to Bombay and Calcutta in search of factory employment.

[21] IOL&R, L/PWD/3/218, Madras RR Letters, collection to Letter 2, 13 February 1864, pp. 15–16. The writer was W.G. Smart, CE of the MR.

harvesting, was to the village economy. Their involvement in railway construction was ephemeral, both in the nature of their participation and in the fact that railway building in any particular locality soon ended. A temporary wage connection was not sufficient to enable railway builders to command the labour of such people. Many temporary construction workers remained, to paraphrase Frere, serfs of those dominant in the countryside. Only government authorities could partially (and they, too, only temporarily) override those who were dominant locally.

Indeed, especially in the 1850s and 60s, there is a refrain of request and protest from the railway companies to the government: use some of your powers to assist us in obtaining labour or, conversely, stop syphoning labour off to PWD projects using your 'special' ability to obtain and retain labour; an ability which one railway company official referred to as 'that influence which the Government always has in India in obtaining labour; no doubt the Government have powers and facilities at times in procuring labour which a private company has not . . .'.[22] An MR Chief Engineer hoped labour would not be diverted to a government-built anicut 'as the tacit power of Government will sometimes divert labour out of the channel it would flow naturally, if free'.[23] Sometimes it was not tacit at all; government officials did impress labour.[24] In at least one instance, railway people impressed coolies and carts 'under the veil of Government authority' with the contractor's servants believing they were employed in government work.[25] This example came from the EIR construction in 1855, but the Deputy Magistrate of sub-division Barh, who attempted to dispel the belief that people had to work as poorly-remunerated railway labour when they could make more by working for a share of the harvest, got little support from the Government of Bengal. His evidence was held to be weak, his action likely to be mischievous, and although he was to prevent any real oppression, he was to help the railway people so long as

[22] *PP (Commons), 1857–8*, cmnd. 416, 1858, *Report from the Select Committee on East India (Railways)*, p. 189. The official was Thomas Walker, Managing Director of the MR.

[23] Selections from the Records of the Madras Government. no. 18: *Report of the Railway Department for 1854* (Madras, 1855), p. 5.

[24] This was not an infrequent complaint. As late as the mid-1870s a railway engineer in Madras complained of bridge construction disrupted by revenue officials of two nearby villages impressing 'a number of his coolies, wudders, and bandies to work on the repairs and drainage of certain roads . . .'. IOL&R, L/PWD/3/219, Madras RR Letters, Enclosures to no. 7 of 1876, dated 16 June 1876.

[25] IOL&R, L/PWD/3/50, Bengal RR Letters, collection 7 to no. 26, dated 8 December 1855.

they behaved properly. One suspects that many in the countryside long continued to view railway construction as a government activity: a belief which may have served the railway builders well. Belief came to correspond to reality when State Railway construction began, though by then the government was less inclined to impress labour (*begar*) for PWD works. However, if we follow Mushtaq Ahmad Kaw in extending begar beyond involuntary and unpaid labour to encompass 'inadequately paid labour undertaken by workers under some form of substantial economic pressure and coercion', then we probably would find a good deal of it in railway construction—though at the price of robbing the concept of begar of much of its useful specificity.[26]

Throughout the half-century, the railways employed village labour for building. But throughout the entire period, and throughout the length and breadth of India, the same complaint was repeated over and over: these people were an uncertain and unreliable source of labour, particularly at harvest time. The dictates of India's climate made agriculture (especially at harvest) and railway construction compete for dry-season labour. Near Berhampore in Bengal, in March 1858, most coolies ran 'away to cut the harvest—inspite of their pay being raised, which of course retained some—but the ancient custom of allowing coolies the gleanings for themselves after the crop is cut is the cause they are so fond of harvesting. For they always make a good thing out of the gleanings.'[27] Another engineer in the nearby Keul Division of the EIR wrote to his CE that there were several seasons in the year 'when every man, woman, and child quit the works for three weeks at a time to gather the crops, and no reasonable pay could detain them'.[28] The engineer pleaded for '*a system being organised to supply us with labour from a distance*' (emphasis in the original) without which the work could not be vigorously prosecuted.[29] In Punjab, in April 1860, the station works at Lahore and the progress of the line south to Multan were slowed because coolies had left for the harvest.[30] In 1862, on the Calcutta and South Eastern Railway the harvest made local labour hard to get, but once the crops were in, the engineer

[26] Mushtaq Ahmad Kaw, 'Some aspects of begar in Kashmir in the sixteenth to eighteenth centuries', *IESHR*, 27: 4 (October–December 1990), p. 465. Indeed, one suspects most construction labour fitted Kaw's definition of inadequately paid work, subjected to some forms of compulsion.

[27] IOL&R, MSS Eur. B. 212, Carrington to parents, 14 March 1858.

[28] BM, IS 180/2, EIR. Report of George Turnbull, CE, 18 August 1858, p. 14.

[29] Ibid.

[30] IOL&R, L/PWD/3/59, Bengal RR Letters, 1860, vol. 20, no. 54, dated 18 July 1860, pp. 53–4.

hoped for a great increase in the labour supply.[31] The Jubbulpore extension of the EIR, in 1864, encountered particularly uncertain weather conditions that disrupted the normal patterns of sowing, harvesting and thrashing, among local agriculturalists. The resultant 'irregularity rendered the supply of labour very limited and fluctuating'.[32]

At the opening of the Amballa section of the SP&DR, in November 1868, the contractor, Henfrey, referred to coolie labour in this way: 'The coolie, though fond of money, prefers perfect idleness, and it is frequently necessary to drive him out of his village in the morning to force him to earn a good day's wages on the neighbouring railway works'.[33] 'Unskilled labour is scarcely to be got' at harvest time for the Madras–Chingleput line, was the lament of the SIR Chief Engineer in 1875.[34] In 1884, Glover testified that local labour could only be got for six months of the year when not working in the fields.[35] 'After March we never count upon local labour; it is of little use to us in India.'[36] In 1887, in Gujarat, local labour was undependable and completely unavailable when the cotton crop was active.[37] The 54-mile Mayavaram to Mutupet meter-gauge line in deep South India was built slowly in the early 1890s because 'the working population is entirely agricultural; labour is consequently available, and that too in a fitful way, in the intervals between cultivating and the reaping of crops'.[38] And, as a final example from the many available, the earthwork for the Tapti Valley Railway in November 1897 was in the same state as in July because of heavy rains and the scarcity of local labour due to harvest operations.[39]

[31] IOL&R, PWD, RR Progs., March 1863, nos. 17–18.

[32] IOL&R, PWD, RR Progs., July 1865, no. 48.

[33] IOL&R, tract vol. 592, *Opening of the Meerut and Umballa Section of the Delhi Railway* (London: W.H. Allen & Co.), pp. 29–30.

[34] IOL&R, L/PWD/3/219, Madras RR Letters, no. 7, dated 16 June 1876, collection 7, p. 33.

[35] *Select Committee on East India Railway Communication*, para. 1915.

[36] Ibid.

[37] IOL&R, P3239, PWD, RR Construction Progs., February 1888, no. 392. This suggests that this construction project depended heavily on locally recruited labour. My argument for the increasing use of circulating labour does not preclude places and instances where the less 'reliable', local labour predominated.

[38] *Railway Report, 1892–3.*

[39] BL, Colindale, *Times of India*, Overland Weekly edition, 20 November 1897, p. 490. The coolies have gone off to the harvest complaint is so common in the sources, especially in the earlier decades, that I stopped noting the point early in my research. The examples in this paragraph are a small selection to illustrate the central concern of the railway builders: the local agricultural population was an unreliable

A discussion of the 'unreliable' workers could be extended greatly. Precisely who they were and why they remained attached to their villages would take us into the complex world of free and bonded labour which, as Pouchpedass correctly observes, was never a clean distinction for many agricultural labourers.[40] Such a discussion would certainly expose as false the expectations of Frere and others that the railways (and other 'blessings' of the Raj) would automatically bring the widespread commodification of labour to rural India and give to individuals the ability to sell their labour power in the best available market. The railways did not do this completely any more than did other British innovations. Ludden, for example, has shown us how British court decisions after 1820 stripped low-caste Pallas of their long-standing rights to fixed shares in the village harvest in return for their labour. These decisions, in theory, turned Palla labour into a commodity but, as Ludden notes, bondage remained and 'the courts could not make labour free to move and seek the market value of its service'.[41] One could go on, as Gyan Prakash has done,[42] to investigate within the Indian context the discourse wherein capitalism has been privileged by its assumed association with free labour and humanitarianism—an association certainly accepted by Frere and, to an extent, by Marx.

In short, the group labelled by the British 'an unreliable source for railway construction labour' represents a large iceberg, the tip of which has been exposed in the discussion above. Another big issue with respect to the railways is—to what extent did railway construction contribute to the proletarianization of nineteenth-century rural India? But, the specific issue in relation to this book is—how did the British get the railways built, and from what sources did they get the labour necessary to do so? One should, therefore, having noted the presence of agricultural labourers in the construction workforce, go on to focus on the group that the British considered more 'reliable' and upon whom the construction process increasingly depended.

There were other groups in rural India who were not attached to village

---

source of labour. If my argument is correct, there were less complaints in later decades precisely because the need for local labour had been reduced by circulating labour.

[40] Jacques Pouchepadass, 'The Market for Agricultural Labour in Colonial North Bihar 1860–1920', in M. Holmstrom (ed.), *Work for Wages in South Asia* (New Delhi: Manohar, 1990), pp. 13–14.

[41] David Ludden, *Peasant History in South India* (Princeton: PUP, 1985), p. 175.

[42] Gyan Prakash, *Bonded Histories: Genealogies of Labour Servitude in Colonial India* (Cambridge: CUP, 1990).

life in a fashion that demanded their presence at certain times of the year. Yet other groups had no home villages at all. These groups plus those from the villages who were inducted into the continuing life of the construction worker provided a source of labour that engineers and contractors came to value and consider more dependable. My sources suggest that this body of circulating labour increasingly came to form the backbone of the unskilled segment of railway construction workers and that those who comprised this swelling body of workers proved ever more willing and able to move regionally and then inter-regionally in search of construction work. Many of these types of workers existed in the pre-railway age. Locally recruited labour never disappeared—and local variations in the extent of its use were considerable—but it became increasingly less important. Two senior railway engineers with considerable Indian experience, LeMesurier and Forde, gave valuable testimony on this issue as early as in 1872. When asked by a committee appointed by the government to examine the feasibility of a railway from Karwar to Gadak: 'Is it not the case that most Indian Railways on this side [western India] are not made with local labour?', both replied that such was largely the case.[43] .

Before further examining the circulating workers, it is useful to provide more information about the physical movement of labour. Since the sources contain intermixed evidence as to the mobility of skilled and unskilled workers, I treat them together in this discussion, even though a specific discussion of skilled workers comes later in this chapter. Four early examples serve to establish the presence at construction sites of mobile labourers and to reveal something of the mechanisms through which they came to the work-sites.

The first comes from Bengal, the Central and North Rajmahal divisions of the EIR, in October and November 1859. The local engineer is relaying to his CE, George Turnbull, the grim and tragic details of a cholera epidemic during which 4000 or more coolies, some 30 per cent of the total, died within a space of about six weeks.

The labourers among whom this epidemic first appeared who were its chief victims came mostly, I believe, from the Burdwan and Bancoorah districts. The first large body arrived in the middle of October; and having been exposed on their journey up to the very heavy rains which fell during the 1st week in October,

[43] IOL&R, L/PWD/3/280, Bombay RR Letters, no. 4 of 1873, Enclosures, *Report of the Committee Appointed by Government on the Projected Karwar to Gadak Railway; and Extensions and Alternative Lines* (Bombay: Government Central Press, 1873). LeMesurier was introduced in chapter three. Forde was a CE of the BB&CIR in the 1850s and later a contractor in India.

and having had to travel over a partially inundated country with no shelter at nights, they were on their arrival in a condition to fall easy victims to such an epidemic as cholera. Large masses continued to arrive almost daily, the utmost exertions of the Engineers failed to get together materials for at once hutting them, and a large proportion had no shelter for many days after their arrival and when cholera was raging among them.[44]

The second and third examples come from the reports of the Chief Engineers, John Brunton and James Berkley, of the SP&DR, Sind branch and the GIPR respectively.

The Province of Scinde contains but a sparse population, which is principally located on the low-lying alluvial land on the banks of the river, where the rich earth yields ample returns with a minimum amount of labour. The Scindee, born and bred on these plains, is naturally indolent and devoid of muscular power; at the same time he is not deficient in talent, easily acquiring a knowledge of account-keeping and writing. The natives of the neighbouring state of Cutch are a much superior race. Cutch send carpenters, masons, smiths, and skilled handicraftsmen to the whole of the northern portion of the Bombay Presidency; and from thence came a large majority of the skilled workmen employed on the Scinde Railway. From the hill tribes of Beloochistan and Affghanistan were obtained a hardy race of labourers; men of great stature and personal strength, but wholly ignorant of the use of other tools than the powrah (a large kind of hoe) and a basket in which to carry the loosened earth.[45]

Nearly one hundred thousand men have been employed upon the Great Indian Peninsula Railway lines at the same time, and as many as forty-two thousand upon the Bhore Ghat Incline, which is 15 3/4 miles in length. This great force has not been collected without considerable trouble; it is not entirely supplied by the local districts, but is gathered from distant sources. Labourers sometimes tramp for work as in England, and on the same work may be seen men from Lucknow, Guzerat, and Sattara. The wants of the works have, however, been supplied by unusual exertions in sending messengers in all directions, and by making advances to muccadums, or gangers, upon a promise to join the work with bodies of men at the proper season.[46]

[44] BM, IS 180/2, EIR. Report of George Turnbull, 18 February 1860, appendix, E.I. Robert to Turnbull, 10 February 1860.

[45] Brunton, *MPICE*, 22 (1862–3), 455–6. The stereotyping present in this quotation deserves a study by itself.

[46] Berkley, *MPICE*, 19 (1859–60), pp. 604–5. In another work there is reference to the workforce on the Bhore Ghat composed of people from 'Bombay, Hyderabad, Gwalior, Goa, Ahmedabad, Kolapore, Broach, Dharwar, Tannah, Sholapore, Rutnagherry, Sattara, Belgaum, Poonah, Sawunt Waree and the Concan'. See IOL&R, L/PWD/2/108, GIPR 1861–2, enclosure to 39 of 1862: 'Annual Report of the GIPR, 1860–1 to 1861–62' (London, 1862), pp. 5–6.

The final quotation comes from George Henfrey, the same contractor quoted above, as to the necessity of driving local village people to work in the morning. This example comes from the construction of the Delhi–Amritsar line of the SP&DR in the later 1860s.

Much less difficulty was experienced in obtaining labour on this contract than on the Eastern Bengal Railway, mainly owing to the Firm having become known to the natives, and to their having established a reputation for fair dealing and punctual payments. Besides the local labour of the Punjaub, a great number of work-people from Bengal, Oudh, and the North-West Provinces, flocked to the line so soon as it was known that the works were fairly commenced: and throughout the execution of the contract a sufficient amount of labour was at all times obtainable.[47]

These four representative examples illustrate the presence, already significant by the end of the 1850s, of considerable movement over substantial distances to railway construction sites. The quotations also suggest two other important ideas. The first is contained in the material from Rajmahal and from the statements of Brunton and Berkley. The idea is that the assembly of these workforces involved considerable effort on the part of the railway builders. Berkley says it most directly—'unusual exertions' were required—but the other two imply the same idea: bodies of workers had to be found, recruited and 'transported' to the work-sites. Some workers in the early years came to railway construction on their own accord or, at least, without the active recruitment of railway builders; many more were sought-out in various ways by railway builders. Initial difficulties in obtaining labour for the EIR's construction from Rajmahal to Delhi, led railway authorities to ask district magistrates to engage labour agents on the EIR's behalf. These agents and their assistants were offered a monthly salary, plus a small commission for each able-bodied coolie, strong woman or lad they recruited.[48]

I suggest that the later 1850s and early 60s were the times when the mechanisms (impersonal, i.e. the labour market, and personal, e.g. the recruiting activities of gangers and agents) of large-scale construction

[47] Sir Arthur Helps, *Life and Labours of Mr. Brassey 1805–70* (London: Bell and Daldy, 1872), p.276. Helps quotes Henfrey in this instance. Henfrey was the agent and resident partner on this particular contract. But if there were no shortages why were some people driven from their villages in the morning to work on the line? Perhaps Henfrey exaggerated for effect when he spoke about coolies preferring perfect idleness; an idea that bespeaks of a constructed colonial discourse about subject peoples.

[48] Leighton L. Appleby, 'Social Change and Railways in North India, *c.* 1845-1914', Ph.D. dissertation (University of Sydney, 1990), p. 91.

labour mobilization were set in place. Another glance at Table 2 reminds us of the magnitude of the construction activity and the consequent demand for labour that existed in those years—all the more impressive, given the fact that railway construction in India did not begin until 1850, move beyond the experimental stage until after 1854, and was severely disrupted in North India in 1856–8. The rise in the demand for railway construction labour was rapid. Moreover, the rise took place alongside an active, government-sponsored programme of public works that competed with the railways for the same kind of labour. The railway builders did have to exert themselves to get the labour they needed. But, the extraordinary demand for labour in that period and the extraordinary efforts to meet that demand, provided the context and the mechanisms for the emergence of a construction workforce which continued to grow throughout the remainder of the century, and the presence of which increasingly freed railway builders from the uncertainties of local supplies of labour.

The quotation from Henfrey illustrates how far the process of construction labour mobilization had proceeded by the later 1860s. The shift to wider, less-constrained labour markets was well under way. Many from NWP, Oudh and Bengal 'flocked to the line', apparently with little effort on the contractors' part other than maintaining their reputation for fair and timely wages, something they had struggled to establish on their EBR contract. A further step is illustrated in the report of the Kutni–Saugor line survey in 1893–4.[49] Once construction started, the surveyor reported, skilled and unskilled labour would come in large numbers from Bundelkhand, Cawnpore and Allahabad. He added:

> In this connection it may be mentioned that, during the reconnaissance, a false rumour got abroad that construction had been started and large gangs of *Loonias*, who had marched down from Banda, Allahabad, Partabgarh and Jaunpur, used to surround the Survey Camps daily, clamouring for work. Some 2,000 or 2,500 of these men were induced to go to the Bhopal–Ujjain Railway, but as many, if not more, returned to their homes.[50]

The end of this process can be found in the words of a twentieth-century engineer. He collected a workforce for the Khyber Railway with little trouble from an 'immense reservoir' of workers in every branch of construction: 'it does not take very long in India, for railway construction is an old-established industry there, and the whisper soon runs through the bazaar that the Sirkar is about to start a great work at this or that

[49] IOL&R, P/4586, PWD, RR Construction Progs., June 1894, no. 180.
[50] Ibid.

place'.[51] Many also applied to be part of the subordinate staff of low-level supervisors, among whom were some whose families had worked on railways for three generations.

The railways themselves contributed to the increased mobility of construction workers as the decades passed. The coolies who struggled from Burdwan to Rajmahal in 1859, did so on foot. Many of the labourers, one suspects, who flocked to Henfrey's works from Oudh and Bengal in 1865–70, travelled at least part of the way by train. By the end of the 1880s, engineers were using the railways to regulate the strength of labour at work-sites and control labour costs. Large gangs of labourers were needed at the Sher Shah bridge during the girder erection phase, 1889–90, but 'at the earliest possible moment the great labour gangs were dispersed, and the work debited with the cost of their railway fares home'.[52]

We know, therefore, that the construction workers on any particular section of line, great bridge or tunnel might have come from considerable distances; that they could have come from many areas and places, although there was a tendency for movement at first to take place within certain broad, regional units: the Frontier, Sind and the Punjab; the gangetic plain extended into the Punjab; western and central India but including also the central Deccan and Sind; South India. We also know that these were not impermeable boundaries and zonal labour markets increasingly overlapped.[53]

Thus, on the Chaman Extension (1887–91) of the IVSR, one found mainly Pathans drawn from Herat, Seistan, Kandahar, Ghazni, Kabul, Jalalabad, Swat and Bannu, but there were also Hazaris, Kashmiris, a few Tibetans, many Punjabis and Hindustanis, plus 'several strong gangs of Mekranis and Mongrel Arabs from the Persian Gulf, with a few Zanzibaris (these Arabs, etc., were most useful on the inclines), Sikh carpenters and masons, and Bengali brick-burners'.[54] Some years later also, on the western frontier, the Mushkaf–Bolan Railway used illiterate Afghani and

[51] Victor Bayley, *Indian Artifex* (London: Robert Hale, 1939), p. 14.

[52] Spring, Technical Paper no. 71, p.20.

[53] Possibly the Malabar coast and other Madras areas roughly south of Trichinopoly may have remained more self-contained. However, petty contractors from Cutch were building railways near Tinnevelly in 1900. At a later time a spatial extension of the labour market that serviced the jute industry of Calcutta also occurred. See Ranajit Das Gupta, pp. 285–309. Unlike urban industries, however, the work-sites of railway construction continually moved.

[54] Technical Paper no. 35: 'Report on the Chaman Extension Railway' by G.P. Rose (May 1894), pp. 3–4.

Indian petty contractors to employ unskilled labourers from Afghanistan, Hazara and the Punjab. 'All skilled labour, such as that of masons, carpenters, and other artisans, was imported from the Punjab and Kurachee at high rates of pay, as the Bolan has a bad name for sickness and expensive living.'[55]

In the early 1890s, railways were built in Assam by Pathan labour which flowed 'freely in, without special recruitment'.[56] Makranis (from the Makran coast beyond Karachi) were also present, as were coolies from much closer—Chapra (west of Patna) and Tirhut.[57] Khols and Santhals from the Central Provinces, and Nuniyas and Sylhetis from Bengal did the earthwork; Nepalese labour did the dry stonework; carpenters, stonemasons and bricklayers came from Punjab, while Cutch supplied masons and bricklayers; riveters and bridge erectors came from Bombay.[58] Also, in the early 1890s, Punjabis and 'Mahratta men' helped to build the bridge over the Kistna at Bezwada.[59] Some years later, in Orissa, a bridge across the Baitarani required brick-moulders and masons who were recruited 'under heavy advances from Lucknow and other large centres in the north-west . . .'.[60] The inclusion into the labour force of people from urban areas reminds us, especially as far as skilled workers were concerned, that railway construction used some non-rural labour.

This information illustrates the fact that the demand for railway construction labour was met within increasingly interlinked, spatially more extensive, labour markets. Some workers, such as the Pathans and Punjabis mentioned above, moved right across India in response to work opportunities; others circulated within smaller markets. Regardless of the size of particular labour markets and their continuing imperfections, their presence, and their increased ability to match the demand–supply equations for labour at particular work-sites, was a significant result of the railway construction process. Railway construction alone did not create these integrated labour markets, but the large demand for labour, the steps taken to meet that demand, and the subsequent ability of the operating railways to transport labour are a major part of the explanation. The

[55] Ramsay, *MPICE*, 128 (1897), p. 254. Plate 2 in this book shows a group of Pathan workers in the 1880s.

[56] IOL&R, P/5003, PWD, RR Construction Progs., July 1896, no. 234.

[57] Ibid., no. 244.

[58] Nolan, *MPICE*, 178 (1909), p. 322. Note the horizontal segmentation of the workforce by region and language.

[59] Spring, Technical Paper no. 71, p. 31. Bombay men had worked on the Sher Shah bridge.

[60] Beckett, *MPICE*, 145 (1901), p. 278.

presence of increasingly effective labour markets in turn made it possible for railways to be built more expeditiously and economically.

What else is known about these groups of circulating labour? We know that most of these workers, particularly the earthworkers, were from the lower margins of Indian society. But can we be more specific? Sometimes the answer is yes; sometimes we know who they were but little more; in a few revealing and important cases we know who they were and something about them.[61] We begin with a discussion of the unskilled workers who formed the majority of the railway construction workers.

Some unskilled railway construction workers came from the tribal populations of India and neighbouring frontier areas. Construction in the western and north-western frontier areas drew upon the tribes of Afghanistan, Baluchistan and the frontier areas of Sind and Punjab, and members of the same tribes penetrated deep into India proper and beyond (Assam) to work on railway construction.[62] Santhals, the Dhargurs 'and the hill-tribes' helped to build the EIR main line in parts of Bengal in the late 1850s and early 60s.[63] After an inspection tour in 1860 of three principal works of the EIR—the Monghyr tunnel and the Soane and Kurumnasa bridges—Governor-General Canning wrote to the Secretary of State for India, Sir Charles Wood: 'Labourers come in freely to seek work, even from the wild Santhal Hills, and no impediment being offered to their return to their homes whenever fancy seizes them, they work well and cheerfully.'[64] Santhals, Poojarsa Nyas, Ghatwals and Moholees worked on the EIR Chord Line in the area of the Nargunjoo Pass in the late 1860s.[65] Tribals were a prominent part of the massive workforce assembled at the Bhore Ghat. Mahars, Minas, Mhangs, Maugs, Kaikarees and Ramosees are mentioned in the sources.[66]

[61] Anand A. Yang, 'Peasants on the Move: A Study of Internal Migration in India', *Journal of Interdisciplinary History*, 10:1 (Summer 1979), pp. 37–58, is one of the few studies to examine push-pull factors in labour migration at the district level. Migrants from the district he examined, Saran in Bengal, did construction work on the Assam–Bihar and Northern Bengal Railways.

[62] Capt. Scott-Moncrieff, 'The Frontier Railways of India', *Professional Papers of the Corps of Royal Engineers. Royal Engineers Institute. Occasional Papers*, vol. 11: 1885 (Chatham, 1887), pp. 213–56; Lambert, *MPICE*, 54 (1877–8), p. 83; Brunton, *MPICE*, 22 (1862–3), pp. 455–6.

[63] *PP(Commons) 1857–8, Report on East India (Railways)*, p. 241; also BM, I.S. 180/2, EIR. Report of Turnbull, 18 February 1860, appendix.

[64] IOL&R, Bengal RR Letters, 23 January 1861.

[65] 'East Indian Railway—Chord Line. The Nargunjoo Pass', *The Engineers' Journal*, 15 February 1869, p. 24.

[66] MSA, PWD (Railway), 1859, vol. 25, compilation no. 215: 'Bhore Ghat.

The tribal groups who first encountered capitalist wage labour when, suddenly, railway construction began in their refuge areas, command attention. Their response to this situation reveals dramatically the beginnings of a major change which continued for decades. Such people, and they were a minority among the railway workers, had lived in a context largely untouched by capitalism of any variety. Waged labour of any sort was unknown to them. According to one observer of the construction of the EIR Chord Line, they had to be broken into the work through good wages (and immediately, piece-work was the managerial preference) and fair treatment; 'undisciplined savages' were converted into 'industrious workmen'.[67] It is unlikely that the people so depicted were either savage or undisciplined, but they were not people familiar with the demands of wage labour. Their 'horror of continuous labour' was only slowly overcome. These tribals illustrate starkly what was also true of all railway construction workers: their own background, their history, significantly influenced the way in which they came to railway labour; there lay the sources of their agency. For some the transition was easy; it wasn't so for the isolated tribal groups.

Probably there were other tribals at the Bhore Ghat and elsewhere, but railway engineers and contractors were little interested in recording the details of the tribal or caste composition of their workforces. Caste or tribal designations were usually more generic, to be mentioned in passing to indicate roughly where the workers came from, rather than to detail their social background. Only when bodies of workers were believed by the British to be troublesome—turbulent or law-breaking—were more specific designations provided. Thus, we know something about the presence of tribal groups on the Bhore Ghat and also that Pariahs were at work at the Vellore bridge works in 1878, because thirty of them assaulted and killed a European supervisor.[68] However, the one instance where I have seen a detailed breakdown of the tribe/caste composition of a construction workforce, unfortunately making canals rather than building railways, supports the position, if such a point needs much support, that most of the unskilled labour came from the lower margins of Indian society. Moreover, the figures from the Bulandshahr branch of

Disturbances amongst the Workmen Employed' and 1862, vol. 36, compilation no. 227: 'Robberies in the Vicinity of the Railway Works on the Ghats'. Also Proceedings of the Legislative Council of India for January to December 1859. vol. 5 (Calcutta, 1859), p. 218.

[67] See fn. 65.

[68] *Madras Times*, 19 March 1878, p. 2.

the Ganges canal in May 1861, were produced in the context of a drought which drove the major agricultural tribe of the area, the Goojurs, to the canal project. These were people 'who have never perhaps in their lives taken a Phawrah in their hands, are now, with their wives and families, performing daily a very hard day's work'.[69] Of the 7363 workers, 3493 (47 per cent) were Chamars and 1797 (24 per cent) were Goojurs; another 363 (5 per cent) were Bhangies. On the neighbouring Fatehgarh branch, the Chamars, labelled 'the working classes' by the engineer in charge, formed 56 per cent of those employed.[70]

There were, however, other groups of people—again from the lower margins—who played a prominent role in the construction of the railways: those whose occupation before the railway age, for centuries or longer, had involved earthwork in such forms as the digging of tanks, wells and canals, or in the making of roads. For these people railway construction offered opportunities to do on an expanded and more dispersed basis, work which they had always done. Evidence suggests that they seized the opportunities, while the railway builders were happy to utilize their services, for in these groups the British found the closest Indian parallel to the navvy. These groups, when available, formed the most effective excavators at a work-site.

The excavators mentioned most often in the records were a tribe (or a caste depending on how one wishes to designate them) called Wudders, Wadders, Wodders or Wodars by British engineers. A more formal transliteration would be Oddar in Tamil, Odde in Telegu and Vadda or Vodda in Kannada.[71] The Wudders of the Deccan do not appear to have worked as agricultural labourers while their activities as tank-diggers can be documented at least back to the Vijayanagar times.[72] The 1871 census called them Oddars or Wuddava and labelled them an aboriginal tribe who, as gangs, took earthwork contracts in which everyone except the very young and the very old took part.[73] The women carried the earth in baskets while the men wielded the pick and spade. 'They are employed largely in

[69] IOL&R, P/190/64, PWD, Agricultural Progs., 23 August 1861, no. 31. Famine relief construction of railways also undoubtedly attracted members of higher status castes and tribes. Desperation is a great leveller. The percentage of Goojurs would have been much lower if famine conditions had not prevailed. Chamars and Bhangis are a part of that group of low castes sometimes labelled 'untouchables'.

[70] Ibid.

[71] F. Richards, *Salem* (Madras District Gazetteers) , vol. I, part I (Madras: Government Press, 1918), p. 187.

[72] Burton Stein, personal communication.

[73] *Census of the Madras Presidency, 1871*, vol. I: *Report* (Madras, 1874), p. 157.

the Public Works Department, and in the construction and maintenance of railways.'[74] A North Arcot district manual called them the Wodda, 'the navvies of the country, quarrying stone, sinking wells, constructing tank bunds, and executing other kinds of earthwork more rapidly than any other class, so that they have almost got a monopoly of the trade'.[75] Thurston's large compendium of South India's castes and tribes contains an extended discussion of the Wudders.[76] He suggests they are Telegu people whose original home was in Orissa and he presents some of the stories then current among the Wudders to explain their origin and occupation.[77] Various sub-divisions of the tribe are listed, of which the two most important are the Kallu or stone-working Wudders, and the Mannu or earthworkers who, in Mysore at least, by the end of the century, did not intermarry or even socially intermix—the stone-workers feeling themselves superior. The earthworkers did lead a more itinerant existence and moved from job site to job site near to which they established temporary dwellings of a distinctive conical or beehive form. The stone Wudders, it appears, had home villages though they would move to a construction site for the duration of a contract. The Wudders, or at least the earthworker division, had a reputation for hard drinking and extended merrymaking once they got a little bit of cash together, and in that respect

[74] Ibid.

[75] Arthur F. Cox, compiler, *A Manual of the North Arcot District in the Presidency of Madras* (Madras: Government Press, 1881), p. 298. Quarrying and well-sinking did involve special skills thus pointing again to the problems involved in an overly facile distinction between those who were skilled and those who were not.

[76] Edgar Thurston, *Castes and Tribes of South India,* vol. 5 (Madras: Government Press, 1909), pp. 422–36. Thurston's material comes largely from the census *Reports* and the district gazetteers; he adds little beyond what one finds there, but he collects the information handily in one spot. H. V. Nanjundayya and L.K. Ananthakrishna Iyer, *The Mysore Tribes and Castes,* vol. 4 (Mysore: Mysore University, 1931), pp. 659–77 has a useful description of the Vodda (Wudders) with photographs.

[77] Francis Buchanan, *A Journey From Madras Through the Countries of Mysore, Canara, and Malabar,* vol. I (London: 1807), pp. 310–13 has a description of the 'Woddas, or Woddaru . . . a tribe of Telinga origin' who retain that language despite being scattered throughout Tamil and Kannada areas. He says they built roads, tanks, wells and canals and also traded in salt and grain. They followed armies to supply grain and in peace transported grain, salt, 'Jagory' and tamarinds. Some were farmers but they never hired themselves out as '*Batigaru,* or servants employed in agriculture'. In short, they were not village bonded. The old and infirm were relatively stationary but the 'vigorous youth' travelled about with their children 'in pursuit of trade'. Tipu Sultan, Buchanan claimed, impressed many Wudders 'to work at his forts without adequate pay', so many left for 'other countries'.

they also paralleled the British navvy. In some districts they came to have a reputation for larceny.

In South India, Wudders appear in the railway records from the start of railway construction. The MR's Chief Engineer, Bruce, wrote of the operations in 1853: 'The earthwork has been performed almost entirely by Wodders—people who work in small gangs, under the guidance of a superior, or maistry, who, being possessed of some trifling capital is raised to that position, as the man of most influence, who can make small advances to the individual labourer, negociate with the employer, and take as his share a certain percentage on the amount paid.'[78] The Wudders remained an ubiquitous and numerous presence among the railway construction workers in South India and beyond. Their presence was mentioned in district 10 of the MR in 1857, and in district 6 of the northwest line of the MR in 1863.[79] Also in 1863 'the Wuddaree tribe, the most useful and efficient class of labourers on earthworks' were stated to be scattered over the proposed route of the GIPR extension, Sholapur to Hyderabad, to join up with the MR northwest line.[80] Wudders, 'professional navvies' were doing most of the railway construction work near Guntakul in 1885, and in Mysore state in the same year delay in starting the line from Gubbi to Tiptur threatened to disperse 'large and useful gangs of Waddahs, i.e. professional navvies or excavators'.[81]

The Wudders attracted the notice of the engineers. Not only were they expert navvies who turned out 'within a given time more hard work than any other labouring class', but their life-style and working methods, especially those of the stone Wudders, were sufficiently distinctive to cause some engineers to describe them. Going evocatively describes them in the period 1867–75. He tells us the Wudders were the navvies of India who were divided into those who worked in stone and those 'skilled in the manipulation of earth'.[82] The latter conducted 'the greater part of the earthwork in Madras presidency, on the railway, in tanks etc.' . . . and always worked by contract.[83] He describes the former thus:

[78] IOL&R, R.2.II, Madras Railway Reports, 1853 to 1855. Madras, *Report of the Railway Department for 1853* (Madras, 1855), pp. 2–3.

[79] IOL&R, V23/156, Madras Govt., *Report of the Railway Department for 1857*, p. 249; IOL&R, PWD, RR Progs., April 1864, no. 6.

[80] IOL&R, L/PWD/3/273, Bombay RR Letters, no. 53 of 28 October 1863.

[81] IOL&R, PWD, RR Construction Progs, March 1885, no. 491 and November 1885, no. 59.

[82] IOL&R, MSS Eur. C. 378, Going typescript, p.9.

[83] Ibid., p. 10.

The stone wudder is a hardy, sinewy fellow, whose stock in trade consists of a house, which, when on his travels, he transports on the back of his donkey, or else on the head of his wife (no remarkable instance of tyranny, since it consists of nothing but a mat and a few bamboo stays); then he has a heavy crowbar, a few iron wedges, some earthen pots, a dog, and a small stock of rice. He is, in fact, a being a good deal resembling the Irish tinker of times not so ancient but that we can recall the picture. Arrived at his quarry, his first care is to lay in a stock of firewood, which he cuts in the jungle, and removes by means of a peculiar bandy or cart with low wheels of solid timber, drawn by a pair of buffaloes. . . .[84]

Going then goes on to describe in some detail how a Wudder quarried rock. He built a fire on the surface of the rock face—usually at night which was the favoured working time—which burned until the heat expanded the rock which, with a dull bursting sound, separated from the layers below. The loosened rock was then broken into fragments capable of being handled by two men by the primitive expedient of repeatedly dashing down a heavy, round, green-stone boulder. The end result was usefully-sized, square blocks of stone which, however, were more 'to be attributed to the natural tendency of the stone to square fracture than to the skill of the wudder'.[85]

A similar process was described by Francis Spring at the site of the Kistna Bridge construction in the early 1890s, though by then a 14-lb sledge hammer was sometimes used in lieu of a boulder. Spring had 286 stone Wudders of whom most came from 'Bellary direction, but they are to be recruited from many places in Southern India and the Deccan'.[86] Each skilled Wudder was assisted by four common coolies and the entire group got out and despatched 52 wagons—about 5200 cubic feet—of stone daily. The Wudders trained local labourers who then worked on their own account. The Wudders and the local labourers together then produced 78 wagon-loads—7800 cubic feet—per day.

The Wudders appear to have originated in South India and to have been concentrated in the Tamil, Telegu and Kannada speaking areas. Records show that they soon appeared in areas beyond South India. Arthur West employed some on road work in the Koyna Valley, north-east of Bombay, in October, 1851. He referred to them as 'Wadars', a 'distinct caste' who were 'first rate workpeople' and who travelled 'over a large extent of country, quarrying stone, doing earthwork, etc. by piece

[84] Ibid., p. 9.

[85] Ibid.

[86] Spring, Technical Paper no. 71, p. 36. Note Spring's reference to recruitment.

work. They live like gypsies in little portable mat huts'.[87] West reported they could do as much digging work as two or three begaris (impressed labourers?), as the local work people were called. Wudders appear to have been present at the Bhore Ghat works by 1859.[88] They, or at least stone Wudders, were deep into Central India in 1864, where 'Wuddary stone' was in demand for some sections of the Nagpur branch of the GIPR. Stone could not be obtained from quarries near the line so it was 'brought to the works by people of the Wudda tribe, who receive so much per stone from the contractors'.[89] More fascinating and more significant yet, however, was the testimony of the contractor, Thomas Glover, in 1884. Glover, whose contracts were in central, west-central and western India was asked by a member of the Select Committee: 'Is a large proportion of the work done by those navvies, trained men, or by coolies?' He replied: 'We principally depend on those Wuddars, as we call them; a gypsy tribe for our earthwork, etc.. The local labour we have to train in every district throughout India.'[90] The Wudders lived in gangs which moved from place to place and they either came to the work-sites or Glover's agents sent for them when needed.[91] A connection that facilitated the flow of labour had been established.

It is not known whether the Wudders moved beyond the Deccan/ South India in response to the opportunities created by railway construction, or whether earth moving and stone quarrying work earlier in the century on British PWD projects, or opportunities that preceded British rule, brought western and central India into their circulatory ambit. One plausible scenario is that the extraordinary recruiting effort to find labour for the Bhore Ghat incline set the Wudders into more active motion, since we know that the recruiting effort extended to Sholapur and beyond to Hyderabad.[92] Once drawn into the western and central railway work, it is reasonable to assume that the connection between the Wudders and the railway engineers and contractors was maintained.

[87] IOL&R, MSS Eur. D. 1184, West Collection, 'Memoir', part II, entry for 8 November 1851.

[88] Proceedings of the Legislative Council of India for January to December 1859, vol. 5 (Calcutta, 1859), p. 218.

[89] IOL&R, P/191/10, PWD, RR Progs., July 1864, no. 59.

[90] *Select Committee on East India Railway Communication*, para. 2091.

[91] Ibid., para. 2093.

[92] The *Gazetteer of the Bombay Presidency*. vol. 17. part 1: *Poona* (Bombay: Government Central Press, 1885), p. 426 enumerates Vadars at 2677. 'They say they came into the district twenty-five or thirty years ago, but from where they cannot

It is also possible that British contractors and engineers came to use Wudder to refer generically to other circulating earthworking groups. The useful article on 'Beldar'—another generic term for those who worked as masons or navvies—in *The Tribes and Castes of The Central Provinces of India*, has this heading: 'Beldar, Od, Sonkar, Raj, Larhia, Karigar, Matkuda, Chunkar, Munurwar, Thapatkari, Vaddar, Pathrot, Takari'.[93] This article confirms the presence of Wudders (Vaddar) in the Central Provinces and points to their South Indian (Telegu and Tamil) origins. However, the article also shows that a number of groups, in addition to the Wudders, most of whom shared a low-status, tribal, derivation, were also part of the grouping, and were variously designated by tribal names or by names that designated an occupation. Thus, Matkuda, or earth-diggers, were usually Gonds or Pardhans.[94] In total, the earth and stone-working castes numbered some 35,000 in the Central Provinces in 1911.[95]

Speculations aside, what is known is that the Wudders were certainly an important component of the railway construction workforce throughout south, central and western India. Where earthwork was concerned, the Wudders were a significant component of the group the British came to call the navvies of India: navvies who migrated over long distances to work-sites, who moved earth more effectively than any other group and who helped to free the railway companies from labour obtained from village populations whose presence at work-sites at certain times of the agricultural year could not be counted on. We cannot quantify the degree of freedom so achieved, but we do know it was considerable. We have the testimony of Glover to that effect presaged by the observation of Lord Elgin.

The Wudders were but one of many traditional earthworking or stone-working groups of whom representatives could be found in virtually any part of India. The Ods were listed as one such group in the Central

---

tell.' Twenty-five or thirty years would place their arrival in the period of the construction of the Bhore Ghat incline.

[93] R.V. Russell, *The Tribes and Castes of the Central Provinces of India* (1916; reprint ed., Oosterhout: Anthropological Publications, 1969), p. 215. A useful introduction to these and other low castes can be found in Stephen Fuchs, *At the Bottom of Indian Society. The Harijan and other Low Castes* (New Delhi: Munshiram Manoharlal, 1981). Chapter 3 on the 'Semi-Nomadic Castes' discusses Odhs, Wudders and others.

[94] Ibid.

[95] Ibid.

Provinces, and in Rose's standard compendium of Punjab tribes and castes, one finds 'Odh, Ud, Od or Beldar' as an entry. Beldar, Rose stated, was an occupational term derived from the *bel* or mattock with which the beldar worked. The common cooly in the Punjab sometimes dug earth but the Od was the '*professional* navvy' (emphasis in the original), and usually the only group referred to as Beldars.[96]

Rose goes on to describe the Od as a wandering tribe whose Punjab representatives appeared to have western Hindustan and Rajasthan as their ancestral 'home' but who established temporary dwelling-places next to the work-sites.[97] They travelled about in family units seeking work on roads, canals, railways, tank digging or house building which they preferred to undertake on small contracts. Men dug the earth, women carried the earth to the donkeys they always had with them, and the children drove the donkeys to the spoil banks. In some parts of the Punjab—notably the Salt Range—they also quarried and carried stone. Colonel Medley refers to the 'Uds' in his treatise on Indian engineering as 'a class of excavators well known in the upper provinces of India, who wander about wherever they can get work and pasture for their cattle'.[98] He, too, refers to their use of donkeys to carry earth.

Enthoven, in his magisterial survey of the tribes and castes of Bombay, discusses 'Ods, Vaddas or Beldars', also spelt, he says, 'Odde, Wodde, Waddar, Vadar and Orh', together as one generic group albeit with various sub-divisions and regional variants.[99] He believed the tribe moved northward from the south, and in the process acquired members from many other castes. 'The skill of the Ods in earth work and masonry has led to a demand for their labour in all parts of India.'[100] In 1901, Ods were to be found throughout the Presidency, including Sind, and they numbered 94,096, according to the census enumeration.

[96] H.A. Rose, compiler, *A Glossary of the Tribes and Castes of the the Punjab and North-West Frontier Province*, vol. 3 (Lahore: 'Civil and Military Gazette' Press, 1914), p. 175.

[97] Ibid. Various contributors to *Panjab Notes and Queries*, vols. I and II (1884) speculated that the Odhs of the Punjab and the Wudders were the same tribe.

[98] Lt. Colonel J. G. Medley, compiler, *The Roorkee Treatise on Civil Engineering in India*, 3rd edn (Roorkee: Thomason College Press, 1877), vol. I, p. 248.

[99] R.E. Enthoven, *The Tribes and Castes of Bombay*, vol. 3 (1922; reprint edn, Delhi: Cosmo Publications, 1975), p. 138, pp. 138–49 provide the most extensive description of the beliefs, practices and divisions of the Ods/Wudders of all the caste compendiums—perhaps because in the series of Bombay District Gazetteers published in the 1880s most volumes have a substantial description of the Wudders.

[100] Ibid., p. 139.

A similar phenomenon was found in the North-Western Provinces and Oudh, where Beldar was a general term for low status, Hindu groups who made their living by earthwork.[101] By the 1890s, in western Bengal and Bihar, Beldar designated 'a wandering Dravidian caste of earthworkers and navvies . . . many of whom are employed in the coal-mines of Raniganj and Barakar'.[102] In Bihar and upper India, Beldar was also a title of the Nunias who were engaged as landless labour in cultivation (some had land), saltpetre-making and earthwork. A later source referred to tribes in eastern India, noted for 'their skill in shifting earth', who were contrasted with ordinary coolies who could not be depended upon because they departed to their fields when agricultural work became urgent. Moreover, unlike the ordinary coolies, the tribes skilled in earthwork did not, it was claimed, work in family units—the men both dug and transported earth.[103] This may have represented either a regional difference present in eastern Bengal or, since the source in question dates from 1913, a shift that had taken place over time.

The Nunias or Luniyas have already appeared in this chapter as the Loonias who flocked in search of work to the Bhopal–Ujjain line while it was under survey.[104] They appeared at railway sites throughout northern India and Assam, where they took earthwork contracts. The men dug and the women and children carried the soil in baskets on their heads. Those Nunias who had no land wandered about in the dry season seeking work, near which they built grass huts for temporary shelter. Descriptions of their activities and ways of living in the 1880s and 90s are mirrored in the books of the twentieth-century railway engineer, Victor Bayley. In a wonderful piece of stereotyping and Lamarckian genetics he labelled them a thieving, criminal tribe with 'a hereditary aptitude for earthwork'![105] Even a nineteenth-century method of piece-work payment

[101] W. Crooke, *The Tribes and Castes of the North-Western Provinces and Oudh*, vol. I (Calcutta: Office of the Superintendent of Government Printing, India, 1896), p. 237.

[102] H.H. Risley, *The Tribes and Castes of Bengal. Ethnographic Glossary*, vol. I (Calcutta: Bengal Secretariat Press, 1892), pp. 86–8. Risley adds: discussion of the beldars gets confused by the fact that the term refers both to an endogamous group and generically to 'the low castes of Hindus employed on earthwork'.

[103] 'The Value of Native Indian Labour', *Railway Gazette*, 21 March 1913, p. 366.

[104] Another aspect of the Nunias is explored in William L. Rowe, 'The New Cauhans: A Caste Mobility Movement in North India', in James Silverberg (ed.), *Social Mobility in the Caste System in India* (The Hague: Mouton, 1968), pp. 66–77.

[105] Victor Bayley, *Nine-Fifteen From Victoria* (London: Robert Hale and Co., 1937), p. 52.

remained unchanged. On Bayley's line, the Nunia contractor stood on top of the rising embankment with a bag of cowrie shells and, as a basket of earth was deposited, he gave 1 cowrie to a child, 2 to a woman and 3 to a man which the contractor then redeemed at the end of the day at the rate of 80 cowries per anna.[106]

What the material presented above demonstrates is that the Wudders and other groups that moved earth and worked in stone prior to the railway age, became a very important source of railway construction labour. Their presence facilitated the construction process; they came readily to the work-sites where they laboured with considerable effectiveness. From the British perspective they were desirable workers and welcomed; indeed, they were sought out. These groups were already accustomed to wage labour under the direction of petty contractors who were usually caste-mates and who probably had extra-economic relationships with them. The task of British-supervised management was more to co-ordinate the work of many gangs of Wudders than to make the Wudders work harder. Why, from the British perspective, fiddle with a good thing, especially since change would have involved more supervision and hence higher management costs.[107] The formal subsumption of such work groups under capital had taken place before the railway age, hence the 'disciplining' to the demands of continuous wage labour like that which occurred to some tribals on the EIR chord line was not necessary.[108]

The unskilled workers were only one part, although by far the largest, of the construction workforce. There were also the skilled workers who comprised some 20 per cent of the overall total and who, at certain localities, might form 50 per cent or more of those employed. Concentrations of skilled workers were to be found at station, bridge and tunnel sites, although the extent of concentration depended, particularly where

[106] A late-nineteenth-century description can be found in Spring, Technical Paper no. 71, p. 54. A similar system using tickets rather than cowries is described in ICE MS. no. 1161, R.W. Graham, 'Description of the Bhore and Thul Ghat Inclines, GIPR', 1866.

[107] A fuller exploration of a similar argument can be found in M.D. Morris, 'Modern Business Organization and Labour Administration. Specific Adaptations to Indian Conditions of Risk and Uncertainty, 1850–1947', *EPW* (6 October 1979), pp. 1680–7.

[108] For example, in building roads under the direction of PWD officials in the earlier colonial period or, more speculatively, when they contracted with Tipu Sultan to build a fort in the eighteenth century or with a landowner or a temple to build a tank in the sixteenth century.

bridges were involved, on the stage of construction—i.e. preparatory works such as approaches and river-training required fewer skilled workers; girder erection required the extensive use of skilled labour.

There is also the question of what constituted a skilled worker. For example, were the masons and stone-quarrying Wudders mentioned above skilled workers? They probably were, and certainly when the British wrote about difficulties in acquiring skilled labour they often had masons in mind. Or, what of brick-makers? Railway building in India generated an enormous demand for bricks; an inadequate supply of bricks was a cause of delay in the building of the EIR in Bengal and the North Western Provinces.[109] But were brick-makers skilled workers or not? One engineer called brick-making a skill, but said it was a skill quickly acquired by even common labourers.[110] There is, therefore, ambiguity in the distinction between skilled and unskilled; some jobs straddled the line, and one can situate the individuals involved in either category.

The skilled workers were, in turn, divided in many different ways. One document from 1855 divided the workforce at one site into carpenter maistries (4 classes), bricklayer maistries (3 classes), stone-cutter maistries (4 classes), builders, miners, ironsmiths, hammermen, lascars and muccadums—plus the presumably unskilled nowganies (lifters and carriers of heavy weights), bellows boys, boys, women (2 classes), bhisties (water carriers) and bullock cart drivers.[111] Berkley referred to 32 different classes of artisans and labourers at work on the Bhore Ghat incline in 1860, of whom 10,822 were drillers (miners) and 2659 were masons.[112] In addition to those already mentioned, this list of 32 included buttiwalas to load and fire blasts, storekeepers, timekeepers, interpreters, filemen, platelayers, trumpeters for mustering workpeople, thatchers and harness makers.[113] Descriptions of work at other sites add vicemen, well-sinkers,

[109] *PP (Commons), 1857–8, Report on East India (Railways)*, pp. 64–5, 238–9. Also Hena Mukherjee, 'The Early History of the East Indian Railway, 1854–1879', Ph.D. dissertation (University of London, SOAS, 1966), ch. v.

[110] *PP (Commons), 1857–8, Report on East India (Railways)*, p.239. The engineer was Colonel W.E. Baker, Consulting Engineer to Government for Railways.

[111] IOL&R, L/PWD/3/253, Bombay RR Letters, no. 51 dated 3 October 1855, item 1794.

[112] James J. Berkley, *Paper on the Thul Ghat Railway Incline: Read at the Bombay Mechanics' Institution, in the Town Hall, on Monday, 10 December 1860* (Bombay: Education Society's Press, 1861), p. 7. He does say quite a bit about the Bhore incline in this article.

[113] ICE, Tracts, vol. 144: *Paper on the Bhor Ghaut Incline*, pp. 47–8.

divers, quarrymen, brick moulders, riveters, erectors and khallassis (sailors, e.g. as riggers) to the list.[114]

Early British sources suggest a relationship between caste and the recruitment of skilled workers; later sources are silent on this issue. One early engineer, when asked if caste extended itself to Indian workers, replied: 'to a minute degree, and those which may be considered the mere manual classes of labour are performed by the lowest caste.'[115] An early contractor said the workmen had 'innumerable and most absurd prejudices' and were divided into castes who would only do a particular kind of work and who would not work with men of another caste.[116] Berkley put it this way:

> Country artisans and skilled labourers have their own methods of doing work, but are capable of improvement and are not averse to change their practice. For operations requiring physical force, the low-caste natives who eat flesh and drink spirits, are the best; but for all the better kinds of workmanship, masonry, bricklaying, carpentry, for instance, the higher castes surpass them. Miners are, on the whole, the best class in the country. The natives strictly observe their caste regulations, yet will readily fall into an organisation upon particular works, to which they will faithfully adhere, and in which they are by no means devoid of interest. Although they cling closely to their gangers, they will attach themselves to those European inspectors who treat them kindly.[117]

The later silence suggests that the British no longer believed caste to be an important factor. Later references to the social origins of Indian workmen tend to stress their region of origin. But can we accept the earlier assessments of the British engineers and officials? Historians are coming to understand better how the British created an ethnographic discourse through which they came to describe and understand Indian society in terms of caste categories—a discourse influenced in turn by the brahmans and other religious literati who 'educated' the British into a certain understanding of Indian society. The result was a distorted comprehension in which 'caste' held too much explanatory power—a power that still acts as a brake upon class-based analyses of colonial India. With respect to the

[114] IOL&R, NWP RR Progs., 23 April 1860, no. 98A; Bell, *MPICE*, 65 (1881), p. 256; Spring, Technical Paper no. 71, pp. 10–27.

[115] *PP (Commons), 1857–8, Report on East India (Railways)*, pp. 64–5, 238–9. Baker's testimony.

[116] IOL&R, Eur MSS. C. 401, Two letters, dated 1851, from Henry Fowler (1821–54), Fowler to Leather dated 2 May 1851, Bombay.

[117] Berkley, *MPICE*, 19 (1859–60), p. 605.

unskilled we are best left with the understanding that these workers came overwhelmingly from the lower reaches of Indian society where people would take up many forms of work in the desperate search for survival. And, when caste or tribal designations are available for fractions and sub-fractions of this class, e.g. the various kinds of Wudders, they designated groups who may have had a particular social identification and identifying customs and practices, but who displayed considerable occupational diversity before and during the railway age.

With respect to skilled workers, there probably was movement from existing artisan castes into comparable work on railway construction. Carpenters, blacksmiths and masons, for example, found many opportunities to practice their trades with the railways during and after construction. Brunton referred to the Cutch carpenters and smiths as 'intelligent and excellent workmen' but who, being wedded to their own work practices, had to be slowly taught by European foremen to work in the European fashion with European tools.[118] Spring, at the Kistna bridge, praised a proposed modification of the gear of the steam dredging gallows as a testimony 'to the high intelligence and great ingenuity of a certain class of Punjab metal workers, the descendents of the old gun and sword makers of pre-British day'.[119] Those members of the Punjabi carpenter, blacksmith, mason and bricklayer castes who became Sikhs and who later formed the composite caste (*zat*) within Sikhism known as Ramgarhias, found the railways to be a fertile outlet for their particular skills.[120] But if carpentry and other skills provided the entrée, Ramgarhias were not subsequently restricted to that vocation. The composite nature of the Ramgarhias and their subsequent movement into other activities such as contracting, alert us to the danger of thinking too much about caste background and too little about changing modes of production when we examine the mobilization of the skilled labour needed to build the railways.

Whether the existing artisan castes provided sufficient manpower to fulfill the railways' needs once extensive construction began, is another matter. Certainly, there were complaints of shortages of some kinds of skilled workers in the early decades; there were also complaints of a want of the requisite skills among the labour that was available. The latter

[118] Brunton, *MPICE*, 22 (1862–3), pp. 466–7.

[119] Spring, Technical Paper no. 71, pp. 52–3.

[120] W.H. McLeod, 'Ahluwalias and Ramgarhias: Two Sikh Castes', *South Asia*, no. 4 (October 1974), esp. pp. 86–9. Also see his *The Evolution of the Sikh Community* (Oxford: OUP,1976), pp.101–2.

problem, however, was solved by on-the-job training. Earlier construction projects thus came to be the training grounds that created the highly skilled workers who were needed then and subsequently for some aspects of railway construction, notably the erection of the iron girders that rapidly became the engineers' medium of choice for spanning India's great rivers.

It is most unlikely that members of existing artisan castes only, or even primarily, monopolized artisan-type work in railway construction. If there ever was a close correspondence between occupation and caste ascription it existed in the social world of individual villages. An important aspect of railway construction is that it created or expanded work communities divorced in considerable measure from the constraints of village life. People of diverse backgrounds could and did move into skilled work that was new to them and sometimes completely new to India. Where skilled work was concerned, railway building created not only work opportunities but also skilling opportunities for some who broke with their vocational past.

The skilled sector of the railway construction workforce exhibited the general pattern of movement to work-sites discussed above. Also, a group of specialist workers appeared, who, trained on the early bridge projects, moved thereafter from one great bridge to another, and whose presence facilitated the progress of work on these bridges. The great bridges of the PNSR, built in the 1870s, employed men who had already been trained in such tasks as rivetting on the bridges built for the Delhi to Amritsar line, 1865–70.[121] In the late 1880s, the girder erection phase of the Sher Shah bridge across the Chenab for the NWR employed some 5000 men; of these, some 1500 were artisans 'of whom a large percentage had been previously employed on one or more of the many large bridges which, during the past 20 years, had been constructed in Northern India'.[122] Many of these same skilled Punjabi workmen then travelled almost a 1000 miles eastward to work on the Kistna bridge for the East Coast State Railway.[123] The initial education of these bridge specialists was undertaken by skilled British workmen who were a more numerous part of the railway construction workforce in the early decades.

Bridge building also created a substantial and continuing demand for masons who were soon recruited to work-sites on an inter-regional basis and from beyond, since we find the occasional reference to the importa-

[121] Mallet, *MPICE*, 54 (1877–8, part IV), p. 69.

[122] Spring, Technical Paper no. 71, p. 10.

[123] Ibid., p. 31.

tion of Chinese masons.[124] We have Brunton's testimony about the use of Cutch masons on the Sind line, while on the bridges of the Jhansi–Manikpur State Railway in 1884, contractors expended considerable sums in expense money to import masons from 'distant places as Sukker, Cutch and Jeypore'.[125] Petty contractors, 'mostly Kutch men', took the masonry contracts for the first 14 miles out of Tinnevelly, for the Travancore branch of the SIR in 1900.[126] Spring mentions the presence of a substantial number of Maratha masons at the Kistna bridge.[127] Masons are the single category of skilled workers, most often mentioned throughout the railway records.

Now that something of the identity of the many segments of the construction workforce has been established, two further questions can be addressed. How were these workers, unskilled and skilled, recruited? How easily were the large numbers of workers needed for the railways assembled—in short, were there labour shortages? Both these questions have been touched upon in the foregoing material but a more focused answer needs to be given to both.

The question of how the workers were recruited takes us back to the discussion of contractors and engineers, and the function of advances in Chapter three. In that chapter it was argued, from the perspective of capital, that advances tied labour to capital and helped integrate the many layers of intermediaries who stood between capital and the people who actually built the railways of India. From the perspective of the workers, the advance filled many functions. It could represent the cash necessary to enable workers to travel to a work-site; it could, if advanced further ahead in time, represent the amount needed to tide a family over the unemployment of a rainy season; it could, at the point where a landless labourer or village servant was first inducted into the life of circulating labour, represent the amount necessary to free the labourer from debt and other bonds to village power-holders; it could represent a mixture of all of these and other functions. The advance usually represented a considerable command of the person who gave it, over the labour of the construction worker. The advances helped to obtain and retain labour.

Workers came to the construction sites in units of varying sizes,

[124] Bray claimed to have imported 30 for his Sind contract. See IOL&R L/PWD/2, register 5, 186 of 1864, letter dated 16 November 1864.

[125] IOL&R, PWD, RR Construction Progs, December 1884, no. 33.

[126] IOL&R, PWD, RR Construction Progs, October 1900, no. 42.

[127] Spring, Technical Paper no. 71, p. 31.

recruited from among the groups identified in this chapter.[128] Gangers—who were either themselves petty contractors, or who sold the labour power of the gang to someone else on a task-work, piece-work, hourly, or daily-rated basis—were the point of articulation between the actual workers and those who supervised the construction process. Bruce, quoted earlier in this chapter with respect to Wudder gangs, discerned the origins and essential role of the gangers. The gangers, variously styled muccadum, sardar or maistry, were the ones who made advances to workers in order to persuade them to come to the work-sites.[129] The same people usually commanded the gangs at the work-sites, although the engineers sometimes tried to enhance their direct control of work by placing men of their choice in charge. The gangers were not always men. On the Bhore Ghat, there is a record of a Mina woman commanding a gang of 33 men (beldars and quarrymen) and another woman who was the muccadum of a gang of 30 women.[130] Gangers stood in different relationships to the members of their gangs. Brunton refers to men working in groups under self-elected muccadums or gangers who made all agreements for work, who received and divided the groups' earnings, and to whom each worker paid a percentage of his wages.[131] In other cases, one can be sure that the relationship was more despotic. One can also be sure that in most instances the gangers had an extra-economic connection with their gangs: connections of common caste membership, of shared kinship, of a common point of origination in the same or a nearby village. It was these extra-economic connections that facilitated the act of recruitment and helped to ensure the security of the advance.

[128] The nineteenth-century railway records do not penetrate to the level of the gangs and the gangers. Only the occasional observation lets us know they were there; that and the heavy weight of the circumstantial evidence. One has to turn to modern observers to obtain first-hand descriptions of the life of circulating gangs and their relationships with their gangers and higher-level recruiters of labour. See Breman, 'Part 1', pp. 41–70 and 'Part 2', pp. 168–209, *JPS*.

[129] Maistry can also mean an artisan, but in this context a maistry was often an artisan who had risen to become an overseer, a charge-hand. Maistry therefore was usually a charge-hand among skilled workers corresponding to the muccadum and sardar among unskilled workers. However, in at least one case I have a reference to the Wudders, 'people who work in small groups under the guidance of a superior or maistry . . .'. Madras. *Report of the Railway Department for 1853*, pp. 2–3.

[130] MSA, PWD(Railwnay), 1859, vol. 25, compilation no. 215, Clowser and Adamson to Berkley, dated 26 January 1859. This probably tells us something about the different nature of tribal and caste groups and it raises the question of the different ways in which different groups may have experienced railway construction work.

[131] Brunton, *MPICE*,22 (1862–3), p.457.

Pre-capitalist elements existed within the emerging capitalist labour market.

The gang leader also performed another important function which added strength to the position. Given the cultural complexity of India, the growth of the construction labour market meant that labourers soon found themselves moving across and working in contexts alien to them, where a different language was spoken, different foods eaten and different social customs practised. They were also exposed to dealing with someone in authority at the construction site, possibly a rough and ready British overseer, likely a former soldier who had taken his discharge in India. The gang leader had to steer his charges through strange social landscapes, to make arrangements for work and to be the bottom link in the chain of authority that directed the work. The more alien the context the more dependent a gang of workers became on its headman or petty contractor.[132]

Many imperfections and immobilities existed in the labour market in the 1850s when railway construction first began (in the teeth of competing demands for labour from PWD projects) and required, in the space of a few years, large amounts of labour in parts of western India, Bengal and the North Western Provinces.[133] 'Unusual exertions', using Berkley's phrase, were needed to obtain the requisite labour. One of Berkley's assistants tells us what these exertions involved in the case of the Bhore Ghat incline:

> There is no local labour & therefore the difficulty of collecting and organizing the workmen had to be commenced almost afresh after each rainy season. This was effected by sending numbers of maistrys or muccadums corresponding to foremen and gangers to the different towns and villages in a circuit of 200 or 300 miles supplied with money to enable them to advance small sums merely sufficient to keep the men on the road. The labourers thus collected were taken to the nearest railway station on the Concan or Deccan where their fares were paid for them to Khandalla or Campoolee at the top or the foot of the Ghat as the case might be.[134]

Apart from the fascinating insight into the extent to which the operating

[132] One study that recognizes this important dimension is Yuji Ichioka, 'Japanese Immigrant Labor Contractors and the Northern Pacific and the Great Northern Railroad Companies, 1898–1907', *Labor History*, 21:3 (Summer 1980), pp. 325–63.

[133] Faviell began operations on his contracts in January of 1856 and by 6 March he was employing some 15,000 people on but 45 miles of line. *The Engineer*, 25 April 1856, p. 233.

[134] Graham, ICE MS. no. 1161.

railways were used almost immediately to move labour, this quotation brings out the effort that went into the initial mobilization of labour and emphasizes the role played by the advance. But soon, for all but the most isolated of constructions, extraordinary efforts on the part of capital were no longer needed. The labour markets became better established and better integrated. Bodies of migratory workers, skilled and unskilled, came to know of employment opportunities. The crucial intermediaries, the emerging petty capitalists, began to link capital and labour across regions and beyond. Established contractors like Glover, or his Indian subordinates, knew how to obtain gangs of Wudders when needed. Even the fresh-faced British Assistant Engineer at his first construction or reconstruction job could, well before the century's end, expect in most circumstances to have petty contractors clamouring for work and assuring the young sahab that the embankments would rise with great rapidity because so many people would be put to work.[135] Numerous petty contractors existed who were like so many ferrets when it came to seeking out work. Fair-dealing engineers also found that petty contractors would become attached to them and follow them from job to job. Likewise, an established engineer in charge of the building of a great bridge could recruit skilled workers from his previous project, perhaps even on an individual basis without advances.[136] However, advances remained a central mechanism for bringing work and the worker together. As the Chief Engineer of the SIR stated at a meeting between officials of the Company and the government in December 1887: 'When labour is scarce at any work, it is almost impossible to collect it from a distance without advances of money being made to men and maistries . . . .'[137]

Advances, however, were also a source of conflict and discord. In one case, the Bray Brothers' contract to build the SP&DR in Sind, the major contractor abandoned the works despite advances to keep him going and left behind some 10,000 to 12,000 unhappy and turbulent workers, among whom were many Afghani tribesmen.[138] More typically, both

[135] A.W.C. Addis, *Practical Hints to Young Engineers Employed on Indian Railways* (London: E. and F.N. Spon, Ltd., 1910), p. 49. Also E. Monson George, pp. 47–9.

[136] Spring stated he could get the best Punjabi labour for the East Coast Railway at Rs 30 to 40 per head, 'but much of it is prepared to pay its own joining expenses, if derived from enterprising districts for the sake of the extra couple of annas of daily wages to be earned in a year or two of steady work'. IOL&R, PWD RR Construction, July 1894, no. 292.

[137] TNSA, Madras PWD (Railway), 23 January 1888, no. 53.

[138] MSA, PWD (Railway) 1859, vol. 60, compilation 327.

Indian and English sub-contractors would abandon their works while still in possession of unfulfilled advances, and in so doing leave the contractor or engineer in the lurch and the workmen unpaid.[139] Other variations included sub-contractors accepting advances, but failing to pay their workmen promptly and/or fully. Again, this generated unrest among the work people and made the subsequent recruitment of labour for those lines more difficult. The petty contractors had to accept the bargains struck in the market place before the labour market could function smoothly.

The workmen, too, sometimes accepted advances and then disappeared from the way-of-works. Henfrey, in the early 1860s, found workmen leaving in groups at night after the attempt was made to withhold the amount advanced from their wages. The coolies then went to other works where they again received advances.[140] Contractors for the Travancore Branch of the SIR, in 1901, sought government help when coolies they had recruited from Cuddapah and brought to the works at considerable expense, absconded.[141] One would like to know more about the causes of these worker departures. Undoubtedly, situation-specific factors were at work. They do, however, highlight the importance of advances as a mechanism which usually facilitated the acquisition and retention of labour, but which sometimes caused discord and departures.

Day labour aside, it appears that most labour which came any distance at all to a work-site was mobilized through the intervention of someone like a ganger, although this did not necessarily mean, in later decades, that active recruiting away from the construction site took place. Depending on the local circumstance the gangers either had to go out into the villages to recruit, or, where the circulating earthworkers and masons were concerned, there was probably an on-going structure wherein leaders represented workers in their contact with employers. Gangs went to work-sites in search of work, rather than the railway builders going to gangs for recruiting. Gangs, one suspects, were generally socially homogeneous, coming from the same region, most likely the same locality, and probably from the same caste or tribe. Thus, there is reference to the muccadums of the Mhangs on the Bhore Ghat, and a supporting counter example from Sind where 200 Cutch masons struck work because 'the contractor's agent had placed a Jusulmeer Muccadum over them under whom they

[139] For an example see MSA, PWD (Railways) 1859, vol. 27, compilation 401.
[140] IOL&R, L/PWD/3/62, Bengal RR Letters, Letter no. 42, dated 25 June 1863.
[141] TNSA, Madras PWD (Railway), 12 July 1901, no. 1192.

refused to work'.[142] There is also an interesting body of depositions to a Punjab inquiry in 1860, where various men working for the contractor, Muhammad Sultan, refer to themselves as mates having varying numbers of coolies under them. Suggestive is the fact that three of the four deposers identified as mates were Arains, while two others, ordinary coolies, were also Arains. Arains were a cultivating-market gardening caste found in considerable numbers in the Lahore district where the construction under investigation was taking place. Arains were also usually Moslems as was Sultan.[143]

The way in which the day labourers were recruited is more difficult to establish. They were, first of all, usually from villages close to the line-of-works. They were, if my argument earlier in the chapter is correct, usually labourers and not excavators or skilled workmen. They were casual labour in the fullest sense of the word: hired daily, and usually paid daily. Addis defined them as 'labour employed solely under the direction of the Assistant Engineer by one of his subordinates'.[144] Did they simply show up in the morning in response to work opportunities? Henfrey talked about the difficulties in getting them to come to work. Other engineers wrote about the initial obstacles and opposition they often encountered when trying to tap this source of labour and some of these engineers recognized that there were men of power in the villages, who could turn on the tap if they chose to do so. On the other hand, it is also clear that engineers sometimes personally paid each worker at the end of each working day. Such was the case, for example, on the Nizam's State Railway where, after the failure of the contractor, the line was built departmentally: earthwork by petty contract and rock cuttings by day labour. The day labour was generally paid directly by the executive officers and the certainty and punctuality of payment 'soon brought men to the line, and there is now little difficulty in procuring a few hundred men out of the neighbouring villages whenever they are required', although the bulk of the work was done by imported labour.[145] Direct payment by a British engineer or his subordinate does not negate the possible presence of a gang structure but it does undermine one important source of ganger control—the receipt and distribution of wages.

[142] MSA, PWD (Railway), 1859, vol. 25, compilation 215, and 1859, vol. 60, compilation 327.

[143] IOL&R, Bengal RR Letters, vol. 20, 1860. Letter no. 64 of 1860, appendix 5.

[144] Addis, p. 128.

[145] IOL&R, PWD, Railway Progs., May 1875, nos. 31–40.

Finally, there is the question of the adequacy of labour supply. Was labour for railway construction available in sufficient quantities? There is no single, easy answer. One has to distinguish between unskilled labour and skilled labour, and between the presence of labour and yet the absence of necessary skills, because, at least in the early going, some required skills did not exist among Indian labourers. It is for the latter reason that skilled British workmen were more numerous in the early years, until Indians were taught the necessary skills. It is also clear that at particular times and in particular localities there were labour shortages: sometimes of skilled workmen and sometimes of unskilled. These localized shortages, which were sometimes a failure to retain a once adequate workforce, had local, albeit recurring and similar explanations: out-of-the-way, thinly populated locations that not infrequently were also disease-ridden (e.g. particularly bad malaria tracts); the sudden descent on an otherwise adequately populated work-site of an epidemic disease, usually cholera, that caused the workers to flee and to which they only reluctantly returned; managerial incompetence or malfeasance; the presence of such a substantial demand for labour that even under conditions of completely elastic supply, time was required to assemble the requisite labour; and, of course, the seasonal demands of agriculture with their particular impact on local supplies of labour.

Advances were one marker of the presence of shortages in a particular locality in so far as they meant labour was being recruited from a distance. However, there was also a customary element to advances and a practical need for them, since many workers could not leave their home village without them—they needed money for travel and money to leave behind for the subsistence of their families if the entire family was not travelling to the work-site.[146] Nonetheless, the single most important development, of which advances were one element since they enhanced mobility and control, was the fact that railway construction created regional and then inter-regional labour markets. There were localized shortages, but there was no enduring shortage of labour. The emergence of more widespread labour markets brokered by petty Indian contractors meant that the supply of labour and the labour needs of many work-sites were pooled and the employers benefitted by economizing on labour reserves since, to the

[146] A good insight into this problem is to be found in IOL&R, P/190/77, PWD, General Progs., December 1863, nos. 13–16. The material deals with advances on public works other than railways but it applies equally, I believe, to railway construction.

extent that short-term needs for labour came to dovetail, one construction worker could stand as a reserve for a number of employers.[147] Operating railways enhanced the physical mobility of workers, thus furthering the integration of the labour markets over wider distances.

In conclusion, there was no enduring shortage of labour, although there were localized shortages throughout the period under study.[148] The shortages, moreover, were more common and more widespread during the first fifteen years of railway construction. Apart from the complaints of the engineers, we can also find support for this view in the fact that there was a considerable rise in wage rates after the mutiny. There were a number of causes for the post-mutiny rise but one cause mentioned by many observers was the great demand for railway construction labour.[149] Demand outstripped supply for a period of time in many regions. Moreover, railway builders and PWD engineers competed with one another for the available supply. However, the inadequate supply of labour was also a function of imperfections in the regional labour markets; there were still major immobilities that restricted the flow of labour, one of which was the physical barrier to the rapid movement of labour. Higher wages, however, had the desired effect and more people made themselves available for railway work.[150] Also, people who had been mobilized for one project became available for work on subsequent projects. Thus, work on

[147] Here I follow closely H.C. Pentland, 'The Development of a Capitalistic Labour Market in Canada', *Canadian Journal of Economics and Political Science*, 24:4 (November 1959), p. 450. Pentland goes on: 'When the demand and supply conditions of labour are dependable enough to permit this pooling, the over-head costs of labour can be transferred from individual employers to the market, that is, to the workers themselves and to the community at large.'

[148] There were also localized surpluses, sometimes in quite out-of-the-way places, e.g. the Kandahar State Railway where in 1883 work could not be found for all who flocked to the construction. See IOL&R, PWD, RR Construction Progs., February 1884, no. 88. The same situation occurred in the latter stages of the construction of the Bhore Ghat incline.

[149] BM, Bombay, 'Memorandum by the Commission appointed to collect information on the subject of prices as affecting all classes of Government servants' (1863); NAI, Home Dept., Public Progs., 30 September 1864, no. 145B; IOL&R, SW 119 Supplement, *The Statistical Reporter*, 21 December 1871, pp. 65–80, 22 January 1872, pp. 94–6, 21 February 1872, pp. 97–101.

[150] Did this represent a partial recovery by rural wage labour of the bargaining power C.A. Bayly claims was lost in the earlier nineteenth century? Bayly, *The New Cambridge History of India*, vol. II:1, *Indian Society and the Making of the British Empire*, (Cambridge: CUP, 1988), p. 147.

the eastern Bengal line benefitted from the completion of the work on the EIR in Rajmahal and Birbhum districts.[151] Later contracts on the GIPR were able to draw on men discharged from the nearly completed Bhore Ghat Incline, and, later still, the Nizam's Railway hired men from the recently completed nearby portions of the GIPR south-east extension.[152] As Addis trenchantly observed in his 1910 railway construction manual:

> Imported labour is much more easily dealt with, as regards matters of sanitation, than local labour, as the imported workmen invariable [sic] belong to a class of itinerant workers who leave their homes year after year and follow the call of the contractor—a very sleuth-hound in nosing out fresh construction jobs. Such workers are experienced in the ways of such works, are to be more relied upon, and are prepared for the contingent vicissitudes by previous experience. They have worked for 'sahibs' before and are, therefore, more likely to appreciate measures undertaken to secure their well-being than those who, for the first time, are required to conform to the 'sahib's' ideas.[153]

In addition, railway builders rarely turned to what should have been the response to prolonged labour shortages, namely more capital-intensive forms of construction. Finally, we have the evidence of Tables 1 and 2. Many miles of railway were built and within reasonable periods of time; millions of Indians built those railways. The presence of a large number of workers and the persistence of labour-intensive methods of construction are compelling evidences of the availability, in most cases, of an adequate supply of labour. The fundamental issue was not the availablity of the needed amount of raw labour but the creation of more integrated, wider spread labour markets that could connect a sufficient supply of workers to the demands of capital for particular quantities of labourers at particular work-sites. The labour markets that emerged demonstrated yet again the flexibility present in the advance of capitalism. The relations of production within the railway construction process involved both the wage-connection and the extra-economic relationships between the workers and the gangers/petty contractors who directly commanded their labour.

[151] IOL&R, Bengal Railway Progs., 9 February 1860, no. 32.

[152] MSA, PWD (Railway), 1864, vol. 16, compilation 238; IOL&R, X464, E. Herbert Stone, *The Nizam's State Railway* (London: Murray and Heath, 1876), p.12.

[153] Addis, pp. 119–20. Addis made a distinction between localities where local labour was scarce and localities where it was plentiful, and he implied the greater use of local labour when available. This is consistent with what I argue, namely that as the century wore on railway builders were no longer dependent on local labour, but they still used it when it was advantageous to do so.

CHAPTER 5

Work and Working Conditions

This chapter examines a crucial step in the building of India's railways; the work itself. It looks at the co-ordinated mental and physical activities of many individuals, activities that produced the route surveys, embankments, tunnels, bridges, stations, lengths of rails, etc. which, collectively, formed the basis for the operating railways. In other words, this chapter deals with the point of production. It seeks to describe some of the main forms of work that went into building a railway and to say something about the conditions of life and work at the construction sites. The management of the work processes is also explored, both from the technical and the organizational perspective. The British, we will find, tried to change some forms of work, accepted others, and modified yet others in their continuing effort to find more expeditious and/or economical ways to build the railways of the Raj.

Work

We begin by looking at the work the British did. They were, in various ways and at various levels, vital to the construction process. They provided the conception and the co-ordination that meshed the different forms of labour of hundreds or thousands of Indian workers, spread across hundreds of miles of the right-of-way or concentrated in their tens of thousands to overcome exceptional obstacles, such as the fifteen-mile climb up the Western Ghats at the Bhore Incline. Certainly, railway construction exemplified the separation of conception and execution, of mental and manual work that characterizes some of the labour processes of advanced capitalism; a separation magnified in a colonial situation. The British engineers conceived and managed; the Indians did the actual work. Even the skilled British workmen, who were a more numerous presence in the early decades, were there more to supervise and teach than to engage in manual work.

One of the tasks of supervision was record keeping. At all times, the engineers, be they representatives of the companies, the State railways or government, had to maintain a level of detailed record keeping that was in advance of anything in Europe or the United States. The price of the guarantee was detailed accounting—detailed accounting which, if anything, became more detailed when the State began to build railways.[1] The overall process of building a line of railway was closely watched and audited. This was the responsibility of the British, assisted by Indian clerks, pay-masters and peons.

Also, the British engineers, assisted by a few Indians, carried out two important preliminary phases of railway development, the surveys. The reconnaissance survey was usually conducted by a senior engineer who established one or more general routes a line might take. The detailed survey that followed (sometimes years later when the building of the line was formally sanctioned) was the work of a number of junior engineers, each with his Indian helpers, supervised by a senior engineer. Each junior engineer was given a section of line whose exact alignment had to be surveyed, staked and mapped.[2] The stakes, cairns and other survey markers, along with detailed, sectional survey maps and charts, provided the blueprints for actual construction.

Surveys aside, the role of the Europeans in the construction process was as follows: Firstly, they played a central role in the conception of a line and in the specifications of what had to be done to build it. This was largely mental work, though some of that work, the surveying, took place in the field often in difficult conditions. Other work took place in the offices and draughting rooms of chief engineers where the plans for the line, the estimates, the drawings and specifications for bridges and stations were drawn up and, if the work was to be done under general contract, where many of the details of the formal contract were developed. The design work for a major bridge might be delegated to a well-known consulting engineer in Britain. Secondly, there was the work of large-scale coordination and organization, the essence of which was to ensure the timely presence of adequate supplies of materials and labour at far-flung work-sites. This task was complicated by the fact that few Indian railways were built telescopically, i.e. sequentially from the starting point to the finish-

[1] Government control also became more detailed as defalcations and other financial problems with the Guaranteed Companies surfaced in the late 1860s. See, for example, IOL&R, L/PWD/3/69, Bengal RR Letters, no. 71 dated 29 June 1869.

[2] Survey work is described in H.W. Joyce, *Five Lectures on Indian Railway Construction and One Lecture on Management and Control* (Calcutta: The Bengal Secretariat Book Depot, 1905), pp. 14–23.

ing point with the work crews always at the railheads and the previously completed lines providing transportation for materials; rather, construction proceeded simultaneously at many different and unconnected sites.

Thirdly, there was the work of medium-scale supervision and organization. A British assistant engineer rarely had less than 20 miles under his charge except in the cases of bridges, tunnels or major inclines. This included laying out work, drawing up local estimates (when the work was done departmentally), advancing or paying out money and the accounting thereof. Assistant engineers, under the executive charge of more senior engineers, provided on-site supervision. Beneath them were European overseers and sub-overseers, a layer which became increasingly staffed by Indians as the century progressed. Once the preliminaries were out of the way and the work commenced, the task of the engineer was to make sure that the work was accomplished according to specifications, before full payment was made. When the work was done departmentally, the involvement of the company or government engineer was more detailed and complicated because of the need, for example, to select petty contractors or to recruit and pay day labour.

There was skill and knowledge involved in these various preparatory and supervisory tasks. Estimating rates, for example, involved the manipulation of three variables: (1) the work one man could do in one day; (2) the wages of that man for one day; (3) the contractor's expenses and profit. All three tasks required the engineer to acquire some knowledge of what was customary in a given area. The notion of what work a man could do (or more likely would do) in a day and how much he should be paid for it did not rest on abstract economic calculations. And there was nothing abstract about choosing petty contractors and bargaining with them.

Once the knowledge of rates and work output was established, the calculations, at least in the case of simple earthwork, were straightforward. If an unskilled labourer could dig and carry from borrow-pit to embankment 100 cubic feet per day, and if unskilled labour in the district was getting 4 annas a day, the earthwork rate was Rs 2–8 per 1000 cubic feet, exclusive of tools and supervision, to which 20 per cent would be added if the job was done by petty contract. However, in the case of masonry rates, the engineer had to consider 'the quarrying of the stone or the manufacture of bricks; then the carting of these to the site of the work; the cost of the collection of kunkar or limestone; the cost of coal or the manufacture of charcoal; the cost of outurning the lime and making the mortar; the cost of manufacture of surki or the cost of collection of sand; the cost of doing the masonry itself, etc., etc'.[3] All these factors had to be assessed and blended to form an appropriate rate.

The British technician—engineer, overseer or skilled workman—might also directly intervene in a labour process by way of organizing the work itself, teaching a new skill, or attempting to introduce a new tool. These interventions were made particularly in situations of special difficulty, such as bridges and tunnels, where skilled labour was at work. No doubt a variety of techniques were used to teach the necessary skills and achieve the co-ordinated work required at some points in bridge building. A twentieth-century source refers to an instance of bridge regirdering where a wooden model was used to communicate with illiterate workmen, among whom six different languages were spoken.[4]

The British railway engineer in India in the nineteenth century had to depend heavily on his own resources. There was no body of substantial, experienced contractors in India who could take on general contracts. This situation did improve as the decades passed. Petty contractors and their workmen also became more knowledgeable. Nonetheless, the engineers remained the transfer agents, in so far as the transmission of Victorian railway-building technology to India was concerned. The British assistant engineer had to be a jack of all trades. One writer said that the engineer had to be both a scientific and practical person, a good administrator, experienced, able to speak an Indian language, knowledgeable in the ways of the manufacture of bricks and in selecting appropriate woods from the many available in India, and careful and prompt in his attention to the routines of his office work.[5] After hard-won experience quite a few engineers met many of these demanding criteria.

Most of the manual work was done by Indians. The construction of a railway line involved three major tasks: (1) the formation of the line; (2) ballasting and laying of the permanent way; (3) the building of stations, workshops, etc. and the installation of machinery, such as signalling devices, the telegraph and water towers. These three major tasks, in turn, required a number of critical ancillary operations without which they could not proceed. These ancillary tasks shared a common element: they involved the provision of supplies and materials. Bricks had to be manufactured in great numbers for bridges, viaducts and buildings. Timber had to be obtained from the forests of India (or overseas), cut into

[3] Ibid., p. 29.

[4] Colam, *MPICE*, 227 (1929), p. 256.

[5] E. Monson George, *Railways in India* (London: Effingham Wilson, 1894), p. 40.

sleepers and keys and then treated (creosoted or kyanized) to preserve them from the weather and termites; other timber was cut into different shapes to provide lumber for bridge or building construction. Stone had to be quarried and reduced to manageable shapes and sizes for masonry work, or even further reduced for ballast. Bar or plate iron had to be obtained and fashioned as required at the work-site. Huge, pre-fabricated bridge girders and rails largely came from Britain, but they too had to be transported to the line-of-works.

Of the three major tasks, line formation was the most arduous and time-consuming, and the largest user of labour. This work was the preserve of Indians whose muscle power assisted by the simplest of tools formed the line. The formation level of the line on which the rails were to rest was the finished top of embankments, the bottom of cuttings, the floor of tunnels, or the surfaces of viaducts or bridges. The formation level had to be finished to a gradient appropriate to the tractive power of nineteenth-century locomotives (which increased with time), and preferably close to a ruling gradient which was well within the capacities of the steam engines and which therefore offered operating economies. Curves in most cases had to be kept above a certain radius in order to enable locomotives to navigate them at speed without leaving the track.

Earthwork, be it in the form of cutting or embanking, was the central task in building the formation level. The engineer or overseer had to make sure the work measured up to the required dimensions and that the earthwork was done properly. Clods of earth, for example, could not be above a certain size, else the embankment would not settle properly, and the slopes had to be of a certain gradient, and so on. This work was the preserve of large numbers of Indians employed as unskilled, gang labourers. Other than effectuating a more co-ordinated use of gang labour and enforcing standards with respect to the quantity and quality of the work, the British altered little the way in which earth was moved and shaped. West noted that most of the immense amount of earthwork on the Bhore Incline, which totalled 6,296,061 cubic yards, was carried in head-baskets by coolies who averaged 15 to 20 miles a day, half of which was with loads.[6] Another writer provided a more colourful description of railway earthworking near Bangalore:

There were crowds of boys, girls and women engaged in transporting earth in baskets of truly solar topee dimensions. So many people were crowded on one

[6] J. J. Berkley, *Paper on the Bhor Ghaut Incline,* with an appendix by A. A. West (Bombay: Education Society's Press, 1863), p. 42.

point that in traversing the space between the places of excavation and the place of deposit, they impeded one another considerably. Their to and fro pace was something equal to but not in excess of a sheep driving one, which we have heard described as being neither walking nor standing still.

Davidson, in 1868, wrote that tilt wagons and other labour saving devices were not used in India. Rather, as described in Chapter four, the custom of the country was followed whereby men, women and children brought earth in small baskets on their heads from excavations alongside the line.[8] In 1889, the engineer in charge of the West of India Portuguese railway noted that 'in India, as a rule, little or no plant is used for earth-works, not even a wheel-barrow', while in 1894, E. Monson George said essentially the same thing.[9] MacGeorge, also in 1894, wrote that a distinguishing feature of the formation of the line in India was the continued use of direct, manual methods of earthworking, 'without, as a rule, the intervention of such aids as "tip" wagons and other mechanical appliances . . .'.[10] In Britain, the widespread use of the steam-driven mechanical excavator marked the end of the era of the railway navvy in the 1880s, while in India, manual methods of earthworking continued to be common in the twentieth century.[11]

Bridges, tunnels and the surmounting of major inclines—the latter often involving extensive bridging and tunnelling—represented special cases in line formation. These situations demanded a radical reshaping of the natural world if a line of railway was to be successfully installed. The limitations of one technology, i.e. steam locomotion, established the limits to which the formation of the line had to be made to conform: the

[7] 'Bangalore', *Madras Times*, 22 October 1860.

[8] Edward Davidson, *The Railways of India* (London: E and F .N. Spon, 1868), p. 162.

[9] Sawyer, *MPICE*, 97 (1889), p. 304; E. Monson George, p. 50. On the West of India line, Sawyer introduced light tramways and wagons on the heavier cuttings which, he claimed, made the work faster and more economical.

[10] G. W. Macgeorge, *Ways and Works in India* (Westminster: Archibald Constable and Company, 1894), p. 327.

[11] The encroachment of the steam excavator in Britain is mentioned in Terry Coleman, *The Railway Navvies* (Harmondsworth: Penguin Books, 1968), pp. 53–5. Coleman also states that in the United States the scarcity and cost of labour led early on to the use of mechanical diggers so that navvy came to mean a steam shovel rather than a man. Victor Bayley, *Nine-Fifteen From Victoria* (London: Robert Hale, 1937), p. 54, describes the persistence of manual methods of railway construction in India well into the twentieth century.

limitations of mechanical engineering limited civil engineering. Rivers, it is true, could always be crossed by ferries, by primitive, non-railway bridges or by seasonal bridges or boats—the IVSR did cross the mighty Indus river at Sukkur by ferry until the Lansdowne bridge was opened in 1889—but this solution was cumbersome and considerably slowed the movement of people and goods.

Whereas much of the earthwork for railways was essentially a larger-scale version of what had been done in India for centuries, the railway bridges and tunnels were new. Tunnels were basically unknown on the subcontinent before the railway age; bridges of the dimensions and weight-bearing capacity needed for trains were, likewise, novel. Constructing these, therefore, meant the transfer to India of exogenous technologies; these technologies in turn required the participation of Indian workers in new work processes. New or adapted skills were needed; workers had to co-ordinate their labour in new ways; the rhythm and pace of work itself had to change and/or intensify.

Shift work was an early innovation. For example, three gangs of work people, each doing an eight-hour shift within a continuous 24-hour work period, built the Monghyr tunnel in 1859–60.[12] At the end of a week the gangs rotated shift schedules. In the 1880s, electric lighting made it possible to build bridges round the clock. The girder erection phase of the Dufferin bridge proceeded 'by day and by night, and sometimes it was necessary to work incessantly throughout the twenty-four hours, by having three gangs, each working continuously for eight hours'.[13] The labour process, sometimes with the use of new technologies, was modified in these and other instances.

Tunnels were the less frequent form of difficult construction, so we will concentrate on bridges and only briefly discuss tunnel construction. If Indian railways had, by 1909, 113 large bridges costing Rs 600,000 or more, the number of important tunnels by the same date, but costing 'only' Rs 150,000 or more, was 72.[14] Tunnels were hazardous sites and

[12] *The Engineers' Journal* (2 March 1860), p. 55.

[13] Walton, *MPICE*, 101 (1890), p. 23. The presence of two Gramme dynamo-electric machines at the Empress Bridge site in the late 1870s provided the means for electric lighting but the engineer in charge chose to erect and rivet in daylight because, he claimed, 'with such reckless workmen as Punjabees, nothing but the most serious emergency would justify night-work, and it was never resorted to'. Bell, *MPICE*, 65 (1881), pp. 255–6.

[14] IOL&R, V/27/722/17–18, Indian Railways. 'Statement showing cost and

required workers with special skills. The range of required skills, however, was narrower than that needed for building a great bridge. Blasting and excavation were known to Indian workmen prior to the railway age, and they worked hard in tunnels despite being crowded together in foul atmospheres.[15] Safety precautions, however, were poor, and tunnel blasting, at least on the Ghat inclines, resulted in considerable loss of life. Timbering the tunnel ceilings in treacherous conditions, however, was a skill slowly learned, and as late as during the construction of the Khojak tunnel (1888–91), 65 Welsh miners had to be specially recruited from Britain to deal with the treacherous, water-bearing strata.[16] The tunnels of the Assam–Bengal Railway, built in the 1890s, required the supervision of experienced British tunnellers.[17]

Apart from the preliminaries—the approaches and river training for example—bridge-building involved two major tasks: sinking foundations into the river bed until a secure footing was obtained, and erecting the superstructure upon the foundations in order to carry the permanent way across the river. The peculiarities of the Indian situation dictated a special response to both these tasks. The wide sandy beds and fluctuating water levels of many Indian rivers necessitated very deep foundations. Failure to do so meant scouring at the base of the foundations during periods (which came annually in most cases) when high water flowed at a high velocity. The scouring caused insufficiently deep foundations to collapse.[18] The engineers also learned from early failures that the decision to bridge safely and more economically (shorter bridges) involved the training of rivers so that high water flows were kept within particular channels where scouring could be better countered.

The SP&DR bridges over the Jumna, Sutlej and Beas were completed in 1869, and all three were severely damaged in the first heavy floods they

---

particulars of the important tunnels on the railways in India up to the 31 December 1909 costing not less than Rupees 1,50,000 each'; and ditto of 'some of the large railway bridges in India up to the 31 December 1909 costing not less than Rs. 6,00,000 each'.

[15] Berkley, *MPICE*, 19 (1859–60), p. 608.

[16] P. S. A. Berridge, *Couplings to the Khyber* (New York: Augustus M. Kelley, 1969), pp. 198–9.

[17] IOL&R, P/5003, PWD, RR Construction Progs., July 1896, no. 243.

[18] A useful general discussion of the problems bridge builders faced throughout north India is found in Sir Bradford Leslie, 'Bridges in the Bengal Presidency', Professional Papers of the Corps of Royal Engineers, Royal Engineers Institute, *Occasional Papers*, vol. 16, paper 9 (1890), pp. 223–57.

faced in 1871.[19] In the case of the Beas bridge, a passenger-train dropped into the river through the broken bridge. The BB&CIR bridge over the Nerbudda river failed repeatedly,[20] as did many others to one degree or another. The considerable width of many of India's rivers dictated the extensive use of iron work. Brick or stone arches could only span short distances, therefore their use meant multiple arches, which implied multiple vulnerable foundations—a solution which at some point also became expensive. Smaller arches resting on many foundations also increased the vulnerability to scouring, since the foundations and their abutments impeded the flow of water. India, therefore, became pre-eminently the land of large iron railway-bridges whose ironwork was for many decades largely pre-fabricated in Britain and then assembled and erected at the Indian bridge sites.[21] This, of course, limited the technological and economic benefits India received from railway construction. Pre-fabrication led British engineers unfairly to label their colleagues in India 'Meccano engineers' since they simply had to fasten the girders and struts together at the bridge sites.[22] However, conditions in India and the complexity of some of the bridge designs ensured that erection was a demanding task that stretched the ingenuity of the engineers and the safety of the workers. Plate 5 taken at the Lansdowne bridge construction helps us to understand something of those demands.

Founding the abutments and piers involved sinking brickwork wells or iron cylinders deep into the sandy river beds.[23] A cutting edge or kerb was

[19] IOL&R, P/590, PWD, Railway Progs., appendix to February 1875: 'Report on the large bridges of the S. P and D. Railway'. This retrospect also notes that since the bridges were built by contractors, the records were few, which emphasizes again my indebtedness, for this study, to the record keeping habits of the colonial bureaucracy.

[20] IOL&R, V/27/722/21, *Reports on the Nerbudda Bridge* by Colonel J. S. Trevor; R. E. Consulting Engineer for Railways, C. Curry, Agent B. B. & C. I. Railway and F. Mathew, Chief Resident Engineer, B. B. & C. I. Railway (1868).

[21] Macgeorge, p. 328, states: 'Iron or steel bridges of large average size are imposed by the magnitude and conditions of the numerous great rivers of the country, and India is consequently, in a pre-eminent degree, the land of heavy iron bridgework for railway purposes'

[22] Michael Satow and Ray Desmond, *Railways of the Raj* (New York: New York University Press, 1980), p. 23. Meccano, or erector sets for the US reader, refers to the Meccano sets beloved (in my case disliked, which makes me wonder why I am writing about bridge building) by generations of British children—a kit consisting of small-scale spars, nuts, bolts, pulleys, etc. with which we were encouraged to build bridges, buildings and machines.

[23] Iron piers of a standardized model were used on the BB&CIR. See Lt.-Colonel

made out of timber or iron and laid on the river bed (which in the dry season might well be exposed—if not, a temporary coffer-dam or an artificial island had to be built to protect the work area). Brickwork was laid on top of the kerb to a height of some 12 to 30 feet, and then, if the structure would not sink into the sand on its own, weighted heavily until it had sunk almost completely into the river bed (see plate 4). Another layer of brickwork was then laid on top of the existing one and the whole process was repeated until a well had been sunk to the desired depth, and the top of the well extended a few feet above low water level. In the case of iron piers, additional cylinders were bolted on to the top as the lower cylinders sank further into the river bottom. As the well sank, sand inside the cylinder was scooped or dredged out from the bottom to help overcome the friction that retarded the downward motion of the well. Once the well reached the desired depth, the bottom was sealed with concrete cement which could set in water that tended to flow into the cylinder from under the kerb and whose flow moved the sand, thus also helping the cylinder to sink. The remainder of the well, pumped dry of water, was plugged with ordinary concrete, brickwork or sand. The abutments and piers were then founded upon the tops of these wells. Plates 3 and 4 illustrate some of the devices and procedures used to sink wells.

This method of well-sinking was based on a principle used for centuries in India but much adapted and improved by British engineers who introduced 'an immense variety of ingenious adaptations' and new machinery to sink the wells more speedily to the necessary depths.[24] Wells 80 to 100 feet below low water levels became common, and the Dufferin bridge across the Ganges at Benares had two of its piers sunk to world record depths of 141 and 140 feet.[25]

Shallow wells, like those suitable for some south Indian bridges, were dug by traditional well-sinkers and divers. One man went into the well

---

J.P. Kennedy, 'On the Construction and Erection of Iron Piers and Superstructures for Railway Bridges in Alluvial Districts', Institution of Mechanical Engineers, *Proceedings* (1861), pp. 174–5. Indeed, under Kennedy's management, all components for the BB&CIR bridges were standardized as much as possible which, among other benefits, speeded construction since workers quickly grew familiar with their construction. However, the spindly iron piers proved inadequate.

[24] 'Bridging Alluvial Punjaub Rivers', *Engineering* (28 March 1890), p. 389. The article also states that the growing cost of labour motivated the engineers to use more machinery. Railways construction in India remained labour intensive but this did not preclude the use of more capital intensive work processes in some contexts.

[25] Macgeorge, pp. 329 and 376.

with a small basket, loosened the material at the bottom with his feet, dived under water to fill the basket, and then handed the basket to his companion at the top of the well who provided an empty basket and then discharged the full one. The two men, changing places occasionally, were able to sink a well some 12–14 feet below water level.[26] Deeper wells required the use of the hand-worked Indian jham (dredge) or adaptations thereof, while the still deeper wells came to be excavated by larger, though at first still hand-worked, British designed dredges which, in turn, came to be supplemented or replaced by powered dredges or sand pumps. Dynamite blasts were also used to assist the sinking and in some cases, for example the Koomar and Ishamuttee bridges on the eastern Bengal line, the bottom of the well was excavated by hand after the water from the bottom of the pier was expelled by compressed air—an elaborate system of valves and trapdoors controlled the ingress and egress of air, workmen and baskets of earth to and from the caisson.[27] The professional literature contained extensive on-going discussions of the best methods of well-sinking and suggestions for new or improved dredges or sand pumps.[28] It is clear, from the cross-references in this extensive literature, that the engineers read one another's accounts and tried to benefit from the mistakes and successes of their predecessors.

Sometimes, the wells encountered strata that could only be pierced by hand digging; sometimes, too, the expensive dredging devices would break free and settle in the murky, water-filled bottoms of the wells. In these cases another British innovation was used. Air-supplied divers were sent down to dig or to retrieve the tools. This was dangerous work because the perils of deep-water diving under pressurized conditions were not well understood.[29] The best diver at the Empress bridge (under construction

[26] Stoney, *MPICE*, 29 (1870), p. 384.

[27] Davidson, p. 221.

[28] See, for example, Bell, *MPICE*, 65 (1881), pp. 248–9; Walton, *MPICE*, 101 (1890), p. 17; Thompson, *MPICE*, 122 (1895), pp. 187–90; Eves, *MPICE*, 145 (1901), 292–7; J. F. Strong, 'On the Apparatus Used for Sinking Piers for Iron Railway Bridges in India', Institution of Mechanical Engineers, *Proceedings* (1863), pp. 16–33; and the following in *Professional Papers on Indian Engineering*: 'The Tonse Bridge', 1st series, vol. 3 (1866); 'Sand Pumps for Well Foundations', 1st series, vol. 4 (1867); 'Bull's Hand Dredger', 2nd series, vol. I (1872).

[29] This was not an ignorance unique to India. 14 died and many more endured severe cases of the bends while working on the St. Louis bridge in the United States in the early 1870s. Compressed-air work did not become properly understood and reasonably safe until the early twentieth century. See H. Shirley Smith, 'Bridges and Tunnels', in Charles Singer et al. (ed.) *A History of Technology*, vol. V: *The Late Nineteenth Century* c. *1850 to* c. *1900* (Oxford: Clarendon Press, 1958), pp. 515–16.

1873–8, across the Sutlej river) was crippled for life when he stayed down for fifteen minutes in 115 feet of sludge.[30] European divers were used at first on bridge constructions but Indians were soon persuaded to take to the diving apparatus.[31] 32 Indian divers supervised by one European foreman were employed on the new Papagni bridge built in the early 1890s.[32]

Apart from the compressed air divers, the bridge foundation wells required little by way of new skills among the Indian workers, though they did require submission to the dictates of more intensely co-ordinated labour and, for those in the wells, fortitude in the face of dangerous, confined working conditions. Excavating—by hand or by providing the muscle power for the ropes, pulleys and winches that operated the hand dredges—brick-laying, loading and unloading the top weights which caused the wells to sink, and the quarrying and pitching of stone around the exterior of the wells to protect against scour, were the main tasks. Mechanized dredges, however, did introduce a category of work akin to the heavy equipment operators one finds in modern construction work.

Because of the confined space within, large numbers of workers could not be used at a particular well. Of course, on a large bridge, more than one well was built at a time and the work external to the wells, e.g. stone-pitching, did require many workers. Nonetheless, well-sinking became the preserve of certain bodies of workmen who, having become disciplined to the patterns of co-ordination and confinement on previous projects, moved as gangs from one bridge project to another.[33] This reminds us that the way in which work was organized was as important as the tools used and the skills involved. And, in so far as the engineers were concerned, the great advance with respect to foundations came in the area of mental work. Generations of engineers learned over time how to build secure bridge foundations in India. The first lessons were learned from Indian techniques and tools, followed by lessons acquired through trial and error. Later improvements in techniques and tools were acquired and transmitted through the interplay of experience (again passed on as

[30] Bell, *MPICE*, 65 (1881), p. 251.

[31] IOL&R, L/PWD/3/219, Madras RR Letters, no. 12 of 1876, dated 3 October 1876.

[32] Thompson, *MPICE*, 122 (1895), p. 190.

[33] IOL&R, India, Director of Railway Construction, Technical Paper no. 86, F.T.G. Walton, 'The Construction of the Godavari Bridge at Rajahmundry on the East Coast Railway' (c. 1900).

further refinements took place) and knowledge learned from professional journals, manuals and lectures. The professionalization of engineering proceeded throughout the nineteenth century and with it came the routinization of education, the maintenance of standards, and the ability of civil and mechanical engineers to learn from the experiences of others.[34] By a different route, but to a similar effect, Indian workmen learned new skills which they preserved and transmitted to each other and to their children.

Good foundation construction was crucial to the security of a bridge, but it was the superstructure that required more and more highly skilled workers. As stated above, India became pre-eminently a land where large railway bridges were built of iron and later of steel. Attractive, arched masonry bridges, often styled viaducts, were numerous but restricted mostly to smaller rivers and gullies, or to the south Indian rivers which did not experience the great volume of water that characterized the flood stage of the rivers of the north.[35] Preparing and erecting the iron and steel superstructures of the great bridges involved work processes that were new to the subcontinent. Engineers, supervisors and skilled European workmen at the early bridge projects had to train workers in the requisite skills. Later projects were able to recruit workers trained on the earlier projects, as described in Chapter four.

The site of the construction of a great iron bridge became for a three to five year period, reduced in later decades to two or three years, a small town humming with productive activity. Even before much construction got under way, houses had to be built for the engineers and foremen and huts for the labourers. Workshops for workers in wood and iron, brick and

[34] A general treatment of the growth of the profession of engineering can be found in W.H.G Armytage, *A Social History of Engineering*, 3rd edn (London: Faber and Faber, 1970). Chapters 12–15 and 21 are most relevant to this study. Armytage argues in chapter 15 that Indian conditions and service in India represented a challenge and a stimulant to British engineering generally in the mid-nineteenth century.

[35] The first stretch of the MR, Madras to Goriattum, had stone bridges across the rivers Cortillaur, Poiney and the two channels of the Goriattum. They involved, respectively, 26 arches, 56 arches, 24 arches and five arches. Davidson, p. 346. The 22-stone-arch Dapoorie Viaduct linking the island of Bombay with Thana built in 1854 still carried main-line traffic in the 1980s. See Satow and Desmond, p. 62. However, the early South Indian bridges over the major rivers did tend to have very shallow foundations which subsequently proved unequal to the force of extraordinary floods. See Stoney, *MPICE*, 134 (1898), pp. 66–79.

lime kilns, and brickfields had to be made ready. Plant and materiel had to be collected and stored. Tramways had to be laid, linking stores, workshops and the actual bridge site. Descriptions of the work-site at the Alexandra bridge (1870–6) carrying the PNSR across the Chenab, and that of the Sher Shah bridge (1888–90) carrying the NWR across the Chenab, were very similar.[36] The Alexandra bridge needed 15 miles of tramways 'which were in constant use by night and day during each working season'.[37] Workers with many different skills were present. Apart from the unskilled, masons and bricklayers, there were brickmakers, mechanics to keep the machines in operating condition, carpenters, blacksmiths, rigging and scaffolding experts, erectors and riveters.

More evocative of the concentrations of labour at these bridges is the following account of the 4 million-rupee Empress bridge. The construction colony of Adamwahan became a town of some 6000 inhabitants with two hospitals, a cemetery, market, brickfield, workshops, etc., all enclosed within an embankment to keep the floods at bay. 'Besides native workmen and petty contractors, the staff consisted of six to eight engineers, twelve to fourteen foremen, about twenty European guards, drivers, and fitters, and one hundred native subordinates, clerks and timekeepers, etc.'[38]

In February 1887, Rudyard Kipling, earning his living as a journalist, captured the sights and sounds of nineteenth-century bridge-building in India. He visited the construction site of the Kaisar-i-Hind bridge across the Sutlej, and described what seemed to the uninitiated to be chaos: 'Lines of every gauge—two-foot, metre and broad—rioted over the face of the pure, white sand, between huge dredger-buckets, stored baulks of timber, *chupper*-built villages, piled heaps of warm red concrete-blocks, portable engines and circular saws.' Toiling men swarmed everywhere. Rivetting had started, and a few hundred men, paid by the piece, worked like devils, 'and the very look of their toil, even in the bright sunshine is devilish. Pale flames from the fires for the red hot rivets, spurt out from all parts of the black iron-work where men hang and cluster like bees . . . '. The noise was startling from one hundred yards away but deafening within the girders where it bounded and rebounded. Earlier, in 1886, the piers had been sunk in a hurry to reach a secure depth before the flood

[36] Lambert, *MPICE*, 54 (1877–8), p. 77 and Spring, Technical Paper, no. 71. These two bridges, in turn, were representative of the kind of work activity to be seen at most large bridge construction sites.

[37] Lambert, *MPICE*, 54 (1877–8), p. 81.

[38] Bell, *MPICE*, 65 (1881), p. 257.

waters arrived. 'Men worked in those days by thousands, in the blinding sun glare, and in the choking hot night under the light of flare lamps, building the masonry, dredging and sinking, and sinking and dredging-out.'[39]

Individual skills were enhanced, but so were the patterns of co-ordinated work, and that enabled Indian workers to put enormous girders into place and then secure one to the other. The IVSR's Lansdowne bridge across the Indus at Sukkur (opened 1889) required the erection of 3300 tons of 'the most awkwardly designed steelwork'. Berridge comments further:

> It was almost as though the designer had gone out of his way to test the ability of the erector. Giant derricks, weighing 240 tons each and 230 feet long, made up of parts topping five tons apiece, had to be erected leaning out over the water and at the same time inclining inwards in the plane at right-angles to the line of the bridge. And, as if that was not difficulty enough, horizontal tie girders 123 feet long and weighing 86 tons each had to be put together 180 feet up in the sky! Although cigar-shaped, neither the struts nor the ties actually came to a point at the ends; the joints were rigid riveted connections.[40]

This work with massive beams, illustrated in plate 5, was both difficult and delicate. Twin struts and ties, over 200 and 100 feet in length, composed of many smaller pieces, had to finish up exactly alike. All this was done in conditions that included high temperatures and violent dust storms which blew 'sizeable chunks of timber off crazy scaffolds. . . .'[41] Four men fell to their death from the great heights; two more died when struck by tools falling from similar heights.

Riveting was an example of a new skill that required the co-ordinated work of a small gang: a hammerman, a holder, a forge-man and a number of helpers. The co-ordinated use of a great many gangs was needed in turn to quickly complete the iron work of an entire bridge. Bridges on the EIR's Jubbulpore extension were riveted in 1866–7 by men gathered from Oudh, Delhi and the north-west, among whom many were old hands and

[39] 'The Sutlej Bridge', *Civil and Military Gazette (C&MG)*, 2 March 1887, pp. 2–3. The article is reprinted in Thomas Pinney (ed.), *Kipling's India: Uncollected Sketches 1884–88* (Houndsmills: Basingstoke, 1986) which, based on Kipling scrapbooks at the University of Sussex, is able to attribute many unsigned articles in *C&MG* to Kipling. The Sutlej bridge, transposed to the Ganges, was the basis for Kipling's short story, *The Bridge Builders.*

[40] Berridge, p. 122.

[41] Ibid., p. 124.

expert riveters who could put in 50–100 one-inch rivets per day 'of a quality and neatness not to be surpassed'.[42] Indian technology, however, was adapted on the Jubbulpore bridges for heating the rivets—clay country hearths seated on a conveniently portable one-eighth inch thick, 2 feet square iron plate, were used in preference to English forges. The Ravi bridge (opened 1871) required 7/8 inch rivets closed by hand at the rate of five rupees per one hundred, by men who, for the main part, had riveted on the bridges of the SP&DR's Delhi line.[43] Three hundred gangs of riveters were needed on the Empress bridge (opened 1878).[44] These workers, collected with difficulty from as far as Bombay and Calcutta, often proved unable to knock in the one-inch rivets which were used about 50 per cent of the time. Subsequently, as an example of the job-shifting mentioned in Chapter four, stronger well-sinkers and erectors (who were mainly boatmen by profession) were then trained to do the work with heavier hammers. Even so, the best gangs rarely put in 100 good rivets per day, and the average was only about 30—with a 50 per cent rate of wasted rivets rather than the usual 20 per cent. If the hammermen were bad, the holders-up were worse: 'a man too listless to hold up for two minutes would spend an hour, if unwatched, in caulking a slack rivet to pass it off as sound'.[45] Mr Macpherson, the foreman riveter, eventually solved the performance problem by devising 'an ingenious telescopic dolly to be keyed up to the work by a large cotter'.[46] There were 16 girder spans in the bridge, each 257 feet long and weighing 405 tons; 30,000 service-bolts were needed to fasten the work together prior to riveting; 35,000 rivets were needed to secure each span for a total of 560,000.[47]

By the time the Dufferin bridge across the Ganges at Benares was completed (1887), riveting had become mechanized. Hydraulic riveters—powered by a set of pumps and an accumulator located on each

[42] Middleton Rayne, 'Description of a Method of Launching Girders, practised on the Jubbulpore Line of the East Indian Railway', *Professional Papers on Indian Engineering*, series I, vol. 4 (Roorkee, 1867), p. 328.

[43] Mallet, *MPICE*, 54 (1877–8), pt. IV, p. 69.

[44] Bell, *MPICE*, 65 (1881), pp. 242–58. A more accessible account, based largely on Bell's paper, can be found in Berridge, pp. 76–82.

[45] Bell, *MPICE*, 65 (1881), p. 256.

[46] Ibid.

[47] The task rate for labour only of Rs 9 per one hundred one-inch rivets was high—double the going rate in Britain, so this was one case where Indian labour was not cheaper.

bank—were operated by Indians who, without previous experience with such machines, quickly became experts in their use.[48] This particular labour process strikingly (can one avoid the opportunity for a pun?) illustrated one development in the evolution of railway construction in India whereby the use of new technology, an effective division of labour and better co-operation and co-ordination among the workers continually transformed the process of production: each span was erected more quickly than its predecessor. However, just a few years later numerous gangs riveted the Sher Shah bridge by hand which cautions us against the assumption that technological development followed a smooth, linear path: the engineers always balanced cost and expedient work against innovation.[49]

Bridges, great and small (regardless of whether the superstructure was iron or not), culverts, station buildings and workshops required an astronomical number of bricks and a large number of labourers to make those bricks. For example, in the 17 miles of the Hullohar division of the EIR in the last half of 1858, 2,000,000 bricks were burnt, 4,500,000 were in the kilns ready for firing, and another 7,000,000 were moulded but not kiln-loaded due to lack of labour.[50] Some of the first bricks had been used to construct 50 brick kilns and 16 lime kilns. Kilns and mortar mills could be substantial structures as is illustrated in plates 6 and 7. Establishing an effective brick-making operation, therefore, was one of the first important tasks an engineer undertook as he began to supervise the construction of a bridge, a section of line, or a building. Brick-making exemplified one of the ancillary labour processes that were crucial to railway construction. The heavy demands for good bricks stimulated and rationalized brick-making in India. Traditional Indian processes tended to turn-out bricks that were badly tempered, badly-shaped and often cracked, thus forcing the railway engineers to ensure not only the manufacture but also the quality of large numbers of bricks.[51] A similar amount of care had to be

[48] Walton, *MPICE*, 101 (1890), p. 23. The use of power tools was a characteristic of the construction of the Dufferin bridge. Steam travellers, for example, were used to pull the dredges from the wells.

[49] Spring, Technical Paper no. 71, p. 20.

[50] IOL&R, L/PWD/3/58, Bengal RR Letters, no. 30 of 1859, dated 19 May 1859. 7.5 million bricks were made in the Monghyr division during the same period. The Dufferin bridge required 1,876,289 cubic feet of brickwork; Walton, *MPICE*, 101 (1890), p. 21. Brick manufacture in turn created a large demand for clay, sand or soorkhi, and lime and firewood for the kilns.

[51] Davidson, p. 105.

given to the manufacture of the mortar which was used to lay the bricks. Poor mortar was the cause for many failures of brick and masonry work on the GIPR in the 1860s.

Detailed instructions for brick and mortar making were provided to the engineers and overseers of the BB&CIR in 1858.[52] Appropriate deposits of clay and supplies of water first had to be located near to which some two acres had to be obtained to develop a brickyard capable of turning out 25,000 bricks a day. Two pug mills driven by bullocks, in which the clay was mixed with water and kneaded to a dough-like consistency, and a shed of some 12,000 square feet for the moulders, had to be established. Three kilns, each with a capacity of 100,000 bricks, were needed for a brickyard of this size. The labour force consisted of:

—25 moulders
—25 attendants to carry the bricks from the drying floor—boys paid at the women's earthwork rate were the best
—13 strong men with barrows, paid higher than earthworkers, to remove the bricks from the drying floor
—37 men mixing clay and wheeling it to the pug mills
—13 men or women wheeling clay from the pug mills to the moulders
—20 men to fill the kilns
—20 men to clean and burn the kilns
—5 extras

Other people, often working under petty contracts, were needed to supply the firewood, clay, and to cart the finished bricks to the work-sites. Brass moulds, which again might be supplied under contract to British design and satisfaction, were preferable to traditional Indian wooden moulds, although at first Indian moulders resisted their introduction because of their weight and difficulty in getting the brick to come off the mould. This resistance, however, was overcome because brick-making was piece-work and the brass moulds, once mastered, enabled the moulder to turn out more bricks per day. Put another way, the labourer was co-opted into the creation of more surplus value for the employer.

[52] IOL&R, L/PWD/2/121, Railway Home Correspondence—C. Letters to and from Railway Companies. Register IV—Bombay, Baroda and Central India Railway Company, 1859: 'Memoranda for the Information and Guidance of Those of the Engineering Staff Engaged on Construction', esp. pp. 4–12. In 1910 Addis, pp. 54–60, is similarly detailed on the subject of 'bricks and brickfields' in his instructions to young engineers. Addis, however, was prepared to see bricks provided by contract but he specified that Indian contractors were not to be permitted to lay out their own brickfields.

Precise instructions were given as to how the moulders should sit and work, how the bricks should be stacked, when the hardened bricks should be moved from the shed to the back lanes, and then on to kiln filling, kiln firing, and kiln emptying. The instructions stress precision and the need to maintain 'neatness, order, and strict discipline' in the brickyard.[53] Fines were suggested for failure to follow procedures. The whole discussion, in fact, is redolent with the vocabulary of the discipline of factory work. The overseers had to manage the coolies carefully because if 'every department be not kept to its work, another department will feel it at once . . .'.[54] Tight schedules were to be maintained to meet the daily output of 25,000 bricks. Twenty men took 5 days to fill a kiln ready for firing, the burn took 4 days, cooling 6 days and then the cart-men needed 4 days to empty a kiln—'thus one of these kilns can be fired every 22 days, allowing 2 Sundays and one cleaning-out day'.[55] No doubt the reality was a good deal less tidy, but nonetheless the aim of the BB&CIR's engineers was clear. Brick production was to be an industrial activity: ordered, rationalized, disciplined, with each worker doing his specialized task according to tight, supervised specifications to enable the 'factory' to produce its quota of bricks within the specified time.

Brick and mortar making for great bridges, stations and workshops also provided opportunities for more mechanization. A more capital-intensive approach was possible in a situation where construction at one site was extended over many working seasons and an investment in steam-driven machinery could be recovered and where, moreover, mechanics were available to keep equipment in working order. Plant in working order could also be sold for use on subsequent bridge projects. Large machines were used to provide the clay mixture for the moulders and to make mortar. Bricks, in short, were a major part of Indian railway construction, the manufacture of which, whether departmentally or by contract, was closely controlled by the engineers. As Addis emphatically stated: 'No item of work should receive more attention than brick manufacture'.[56]

The effort to make good bricks was continuous. Different local conditions demanded different approaches and, interestingly, the movement of labour from one project to another could be an obstacle to good brick-making. It took some months before the contractor's agents on the Jubbulpore extension of the EIR got a good brick manufactory going,

[53] 'Memoranda', p. 9.
[54] Ibid.
[55] Ibid., p. 10.
[56] Addis, p. 58.

because 'habits and methods acquired on previous work by sub-agents and foremen' had to be broken.[57] These men could not make bricks at Jubbulpore exactly as they had at Cawnpore or Rajmahal, so they concluded that good brick was impossible because of 'some peculiarity in the soil, fuel or water . . .'.

Brick and mortar manufacture and the laying of bricks was, therefore, an important part of two of the main tasks of railway construction: the formation of the line and the building of stations, workshops and staff housing. Important main-line stations were imposing structures whose external shapes mirrored the changing moods of the rulers of the Raj. Lahore's grim and austere station reflected the uncertainties of the post-mutiny period, while the exuberance and lavishness of Bombay's Victoria Terminus (1887) mirrored the late-Victorian confidence the British felt in India and elsewhere as Queen Victoria, Empress of India, approached her Golden Jubilee. Victoria Terminus was designed by its architect and supervisor of construction, F.W. Stevens, to be a blend of Venetian Gothic and Indo–Islamic styles.[58] The result was a magnificent, luxuriant coup d'oeil. Workshops, if more utilitarian in appearance, were also imposing structures. Workshop complexes covered hundreds of acres, and the adjacent town sites with their neat brick houses for senior staff, many acres more.[59]

Indians, however, had been constructing buildings on a grand scale long before the British came to the subcontinent. Ahsan Jan Qaisar, for example, has described the large concentrations of complexly divided labour, the different pay-scales and the simple yet effective tools present

[57] IOL&R, P/191/11, PWD, RR Progs., November 1864, no. 41.

[58] Satow and Desmond, pp. 33–5 describe, with pictures, the Victoria Terminus. Elsewhere in the book they provide pictures of other stations. Stevens represented another kind of British professional associated with Indian railways, the architect. The successful culmination of Stevens' long struggle with Government to get a sizeable bonus for his work on the Victoria Terminus can be followed in IOL&R, P/3689, PWD, General Progs., January 1890, nos. 80–3.

[59] The Naulakha workshops in Lahore covered 126 acres. The later Moghulpura shops at Lahore, built in the early twentieth century when the Naulakha site proved too small, embraced over 1000 acres. See Kerr, 'The Railway Workshops of Lahore', in S.S. Dulai and A. Helweg (ed.), *Punjab in Perspective. Proceedings of the Research Committee on Punjab Conference, 1987* (East Lansing: Asian Studies Center, Michigan State University, 1991), pp. 67–77. The Jamalpur shops of the East Indian Railway covered 99 acres at the century's end. *Imperial Gazetteer of India*, vol. 14, new edn (Oxford: Clarendon Press, 1908), p. 44. The kind permission of the Railway Board enabled me to visit Jamalpur in February of 1983: many of the nineteenth-century workshop buildings are still in use as is the adjoining town site.

at building construction in Mughal India—construction whose magnificent accomplishments still grace the Indian landscape—based on which, he argues, 'we can easily visualize the complex nature of organizational control and discipline' that must have been present.[60] Brick and mortar making was an important part of Mughal building.[61] The novel element introduced by railway construction was the substantial expansion (within a short period and on an all-India basis) of the demand for many more and better bricks.

The third major element of railway construction was ballasting and the laying of the permanent way (the rails). Bricks played a role in this task as well. Superior forms of ballast—hard, unfriable materials such as broken stone, gravel, pebbles, blast furnace slag or coal cinders—were not always available and recourse had to be taken to inferior substances such as sand or broken-up, hard-burnt brick.[62] Ballasting was a matter of some complexity, since the ballast had to be laid to precise depths and be composed of materials which provided good drainage but did not disintegrate in water; ballast also had to be of a size and hardness which would not crush or break under the impact of the trains, yet not be so light as to be moved or lifted by the passing trains.[63] Considerable labour was involved in ballasting a line from the provision of the ballast itself—often by petty contract from groups like the stone Wudders—to cartage to the line, and the actual laying of the ballast. This work, however, was largely unskilled and, apart from the need to work in co-ordinated gangs, involved little that was novel.

The next step in the laying of the permanent way was to imbed the sleepers in the top layer of the ballast. Few aspects of railway construction in India generated more discussion than sleepers.[64] Should wooden or iron sleepers be used? What kinds of wood were best suited to Indian conditions? How should the wood be treated to increase resistance to the Indian climate and to Indian insects? And if iron pot sleepers were to be

[60] Ahsan Jan Qaisar, *Building Construction in Mughal India* (Delhi: OUP, 1988), p. 13 and chapters 2 and 3.

[61] Ibid., pp. 16–17.

[62] Slag and cinders were usually not procurable in large quantities. Moroever, they contained sulpher which reacted badly upon the rails and the sleepers. See Joyce, p. 33.

[63] Macgeorge, pp. 264–5; Joyce, pp. 32–3.

[64] Summary discussions can be found in Macgeorge, pp. 268–9 and Lt. Colonel J. G. Medley, compiler, *The Roorkee Treatise on Civil Engineering in India*, 3rd edn (Roorkee: Thomason College Press, 1877), vol. II, pp. 339–41. Also see M'Master, *MPICE*, 18 (1858–9), pp. 417–44.

used, what should their shape be? Wood proved to be the most frequent choice, but the provision of adequate supplies of suitable wood was a never-ending problem. Wood was imported from Britain, Europe, Australia and elsewhere, and the forests of India were searched and ravaged for supplies of sleeper wood. Thus, the 10′ × 12″ × 6″ sleeper which was standard for the 5′ 6″ gauge was placed in its bed of ballast after a long chain of operations. Indian wood would be felled in a forest, possibly quite distant, by foresters in the employ of a timber contractor; the logs would then be transported, sawed into the right sizes, inspected, treated and transported again to the construction site. Much of this work was contract work and the labour involved was not novel or complex. The relations in production were more important than the means of production. However, the entire process from forest to installed sleeper involved a substantial number of workers and a complex chain of supply and sequential work operations. And, although it is a point that goes beyond the concerns of this book, it is clear that the demands of the railways for wood—prime wood for sleepers, buildings and carriages, and lesser wood for firewood for kilns and for the fuel of early locomotives—increased the exploitation of India's forests and the pressure on forest-dwelling peoples.[65]

The final step in the laying of the permanent way was the installation of the rails themselves. Plate-laying, as it was known, was a precise operation that required the co-ordinated efforts of a number of skilled groups of workers who, collectively, along with their supporting cast of carpenters and blacksmiths, formed the rail-head gang: a rail-laying squad, the fish-plate men, the spiking squad, the packers, the straightening gang and the finishers.[66] Plate 8 shows a crowded scene as plate-laying took place at the Peshawar City station in the 1880s. The rail-laying squad placed and aligned the rails in the chairs which rested upon the sleepers. The squad also rectified the spacing of the sleepers as needed.[67] The fish-plate gang brought the rails precisely together, allowed for the expansion

[65] The heavy demand for sleeper timber for the EIR Jubbulpore extension, the reckless waste of timber and the potential consequences of both for the forests became controversial as early as 1860. See IOL&R, P/237/29, North Western Provinces, RR Progs., April 1860, nos. 57–60. Also IOL&R, P/217/37, North Western Provinces, PWD, RR Branch, April 1864, nos. 65–71. The pressure on and the resistance of forest peoples can be followed in Ramachandra Guha, *The Unquiet Woods. Ecological Change and Peasant Resistance in the Himalaya* (Berkeley: University of California Press, 1990) and in Richard Tucker, 'Forest Management and Imperial Politics: Thana District, Bombay, 1823–1887', *IESHR,* 16:3 (July-September 1979), pp. 273–300.

[66] Addis, pp. 113–17.

[67] Some varieties of flat-bottom rails did not require chairs.

space, and then bolted the two ends together with the fish-plates, though at this stage only two bolts per joint were used. The spiking squad secured the chairs to the sleepers and wedged the rails into the chairs by means of wooden keys (wedges). The packers covered the sleepers and rail bottoms with ballast or earth. Straighteners made final adjustments to the alignment of the sleepers, and the finishers were a group specially deputed to complete the bolting of the fish-plates. Each sub-gang had approximately 12 men. Approximately, one mile of track required some 1700 sleepers, 115 tons of rails (a rail being 20–4 feet in length), 35 tons of chairs (about 24 lbs each), 5 tons of fish-plates, 3 tons of spikes, 1.5 tons of nuts and bolts, and 1.5 tons of wooden keys.[68] A properly organized and well supervised rail-head gang was able to lay half a mile of track per day. Crossings and points required special attention and took longer.

British plate-layers, along with the other skilled British workmen, were used extensively in the early years of Indian railway construction.[69] However, it was soon found that Indians could be taught to do the job, and during the 1860s plate-laying became a job done largely by Indians under European supervision.[70] In fact, on the MR, which tended to be a leader as far as the training and employment of Indians were concerned, the Chief Engineer wrote, as early as in 1854: 'The progress of the natives in learning to lay rails, under the tuition of Europeans, has been most satisfactory, and I am convinced that, when carefully directed and fairly remunerated, the natives of this country will be found quite able to perform many of those duties for which they are generally considered unfit.'[71] Apart from exposing the European attitude which retarded Indian employment in some railway jobs, CE Bruce correctly identified the future trend.[72] In plate-laying, as in most other aspects of railway construction, Indians could and eventually did perform most of the jobs.

[68] Medley, 3rd edn, vol. II, p. 343 giving the specifications for the EIR, Jubbulpore Branch. The metre-gauge lines used lighter rails, lighter chairs and so on. A major incline required heavier rails and more closely spaced sleepers.

[69] *Railway Report 1859*, p. 30: 'In addition to the Engineering Staff, it has hitherto been found necessary to send out almost all the skilled labour required for the construction and working of the lines.'

[70] M'Master, *MPICE*, 18 (1858–9), p. 421.

[71] IOL&R, W2383, Selections from the Records of the Madras Government, no. 18: *Report of the Railway Department for 1854* (Madras, 1855), appendix one, p. 6; Davidson, p. 100, comments on the leadership of Madras in employing Indians. He, p. 101, expresses some of the British prejudices that delayed Indian employment in some railway positions.

[72] The prejudice was more pronounced and more slowly overcome where skilled operating line jobs like locomotive drivers were concerned.

In 1874, CE Cheyne of the Holkar State Railway was 'much struck with the intelligence shown by Native Foremen plate-layers' who could without supervision 'conduct all ordinary work connected with the road' and among whom were 'more than one who can put in a set of points and crossings very creditably'.[73] This led Cheyne to suggest the next step, namely that with proper training Indian inspectors of plate-laying could replace Europeans at a considerable saving to the State Railways. He then went on to suggest other areas, including engine driving, where Indians could be more extensively employed. The Government of India responded positively to the suggestions and set in motion a process to put them into operation. The prospect of lower labour costs was an effective argument and that argument slowly opened most doors to Indian employment in railway construction, although, as demonstrated in an earlier chapter, the movement into the engineering profession was painfully slow. But, in so far as plate-layers were concerned, by 1885, London was requesting and getting 300 Indian plate-layers to serve the military operations in the Sudan.[74] Plate-laying was a new skill, a skill of the railway age, but, like most railway construction work, it soon became an Indian preserve.

More fundamentally, all this was commodified work for wages. If the establishment of the wage connection represents a formal condition of the subsumption of labour under capital, the same connection is used by employers to attempt to maximize the output of labour. The way in which a wage is earned is an effective tool for the management to establish control over the labour process at the point it counts the most to capital: the realization of surplus-value. It is this, not control for the sake of control, that shapes the conduct of work (although a given employer may believe, perhaps incorrectly, that control per se is necessary—individual employers are no more omniscient than individual workers). Wage structures—bases of payment, standards for payment, frequency of payment, differential pay-scales and premia—offered employers a way to make their control over labour more effective.

Those who employed railway construction labour used the wage connection to enhance the realization of surplus value or, as they would have conceptualized their goal, expeditious, economical construction.

[73] IOL&R, P/558, PWD, RR Estabs. Progs, July 1874, nos 4–. Subsequent responses and counter responses from the senior railway staff and further Government initiatives generated a good many proceedings. The central debate was over operating line employees like drivers and fitters and it is to the history of those employees, rather than the construction workers, that this issue most belongs.

[74] IOL&R, P/2536, PWD, RR Estabs. Progs., June 1885, nos. 101–28.

Piece-work and its close relation, task-work, were the preferred bases for payment.[75] Certainly the BB&CIR brickyards operated on a piece-work basis, which when wedded to a small technological innovation, the brass moulds, increased the production of bricks. The engineers, contractors and petty contractors repeatedly instituted piece-work or task-work, although they recognized that some work processes required greater attention to quality which was more likely to be produced by supervised, time-rated labour. Piece-work provided a seeming terrain of compromise: it co-opted the workers. Workers came to believe that they controlled the tempo of work and that they would benefit from an increased output. As Kipling said of the riveters at the Kaiser-i-Hind bridge: they worked like devils because they were paid by the job.

Since retention of workers was a major concern of employers, at many sites, other forms of worker control could be practised, using different aspects of the wage connection, including advances against wages. Paying irregularly or short, even if it created discontent, could be used to keep workers at a site. Commissariat arrangements, which had to be put in place to retain workers in many isolated localities, could be manipulated to control workers and enhance profit margins.[76] Berkley stated that the employer-owned or controlled 'Tommy Shop' where workers were advanced high-priced provisions against their wages (the truck system), 'commonly discountenanced at home, is beneficial in India'.[77] A fascinating reason for this can be found in a statement by Thomas Brassey: 'The execution of the works on a railway in India is generally undertaken by small contractors or middle men, who in many cases were shopkeepers'.[78]

In summary, we have seen that the Indians employed in railway

[75] I use task-work to refer to a situation where a gang of workers would contract to perform a substantial task, e,g., to cut 1000 cubic feet of earth, which might be finished, more quickly or slowly, as the group paced itself over a number of days. This is a form of piece work but piece work per se is best reserved for the individual or small gang who would rivet, for example, for a payment of so many annas per 100 completed rivets. Piece-work involved a more directly visible relationship between output and payment.

[76] I frequently came upon references to contractors or engineers setting up bazaars or establishing commissariat arrangements but only Berkley provides an inkling into the relationship that prevailed between the workmen and those provisioning facilities.

[77] Berkley, *MPICE*,19 (1859–60), p. 606.

[78] Earl Brassey, *Work and Wages* (London: Longmans, Green and Co., 1916), p. 21. Brassey was the titled (on the basis of his father's accomplishments) son of the great contractor writing a book based on the papers and records of his father. The reference to shopkeepers as small contractors is fascinating and tantalizing.

construction participated in a wide variety of work processes. These processes ranged from activities involving little skill to those requiring considerable skill; from processes that changed little from pre-existing techniques, tools and patterns of work organization to forms that were completely new with the coming of the railway age. All those involved were enmeshed in commodified work for wages but old and new labour processes, the formal and the real subsumption of labour under capital, could and did co-exist at the work-sites.

British engineers and other transfer agents did attempt to alter labour processes through technological innovation and by instituting new forms of organization and new rythms of work. The technological innovations, great and small, were extensive: improved dredges, pneumatic riveters, brass brick-moulds, electric lighting, better and larger lime-kilns and steam-driven machinery. Work was organized and re-organized, supervision tightened, and methods of payment reworked: for example, the re-organization and intensification of brick-making, tighter and more rigorously applied specifications for everything from clods of earth in embankments to rivets in bridge iron-work, the intensely co-ordinated and complexly divided work of plate-laying, shift-work and piece-work. Alterations were made in the way work was conducted, supervised and assessed.

But, and the caveat is crucial, innovation usually came in small, adaptive steps. Railway construction in India spread commodified work and enlarged capitalist labour markets but it did not, in most instances, revolutionize labour processes. The consistent improvement from the 1850s to the 90s, in so far as the work of construction was concerned, was due to the fact that British engineers and related experts learned how to build railways better, more expeditiously and/or more economically within the context of Indian physical and social conditions. They became more knowledgeable engineers and better managers precisely because they adapted to Indian conditions.

The process of learning and adapting is well illustrated by the construction of the Sher Shah and Bezwada bridges in the late 1880s and early 1890s. The two bridges were constructed at sites far removed from one another. The Sher Shah bridge with 17,206-feet spans, carried a line of the North Western Railway over the Chenab river near Multan in Punjab; the Bezwada bridge with 12 spans of 300 feet carried the East Coast State Railway over the Kistna river.[79] The bridges were built departmentally,

[79] The following discussion of these two bridges is based on the splendid, detailed description to be found in Technical Papers no. 71: 'The Bridge of the North

with contractors supplying materials and undertaking some specific jobs.

Both bridges had the same senior engineers. Francis J.E. Spring was the senior executive engineer on the Sher Shah project under the CE, J.R. Bell. Preonath Ghose was another executive engineer.[80] At Bezwada, Spring was the CE and, during its construction, he also assumed charge of the entire East Coast Railway; Ghose had charge of all work outside the bridge itself, including the supply and manufacture of materials. These two men contributed, one can assume, to the movement of skilled workers from the completed Sher Shah bridge to Bezwada (discussed in Chapter four) and to the use of similar river training techniques at both sites.[81]

The construction of these two bridges displayed no disjunctive advance in the technology and techniques of bridge construction. However, the evolutionary advance from the first decades through the Sher Shah and Bezwada bridges and into the twentieth century was cumulatively striking, particularly where the sinking of foundations was concerned.[82] At Sher Shah and Bezwada, less advanced tools than were available were sometimes used. The Sher Shah bridge was hand riveted although

---

Western State Railway over the Chenab at Sher Shah, 17 spans of 206 feet, and the Bridge of the East Coast State Railway over the Kistna at Bezwada, 12 spans of 300 feet', by Francis J.E. Spring.

[80] Ghose had been appointed an apprentice engineer in October 1873. He was first posted to a division of the Lower Ganges canal. By the time the Sher Shah bridge was built he was an executive engineer, fourth grade in the PWD (Spring was a first-grade executive engineer). He previously served on the Ferozpur Bridge works for which his services were acknowledged by the Government of India. Government of India, Public Works Department. *History of Services of the Officers of the Engineer, Accounts, and State Railway Revenue Establishments, including the Military Works under the Military Department*, 4th edn corrected to 31 December 1888 (Calcutta: Superintendent of Government Printing India, 1889).

[81] In 1903 Spring, by then the officiating director of construction for the State Railway system, carried the science of river-training another step forward with his substantial *River Training and Control. Being a Description of the theory and practice of the modern system entitled the Guide Bank System, used in India for the Control and Guidance of Great Alluvial Rivers* (Simla: Government Central Printing Office, 1903). This work is an excellent example of the accumulation, transmission and codification of knowledge that was central to the routinization of the technical side of railway building in India.

[82] The advance is captured well in IOL&R, V/27/722/19, H.W. Joyce, *Economy in Bridge Design and Construction. A Series of Six Lectures Delivered to the Students of the Sibpur Engineering College during March 1914* (Calcutta: Bengal Secretariat Book Depot, 1915).

pneumatic riveters were in use elsewhere in India. The pneumatic riveting plant from the Dufferin and Sukkur (Indus) bridges was used to good effect on the Kistna bridge, as were screw-geared winches and steam pile-drivers for the erection of the temporary bridge. However, at the Kistna bridge, the wells were sunk by bullock-powered dredges. Each dredge required 4 bullocks, one bullock man and 2 men in the well. Sixteen dredges—and thus 64 bullocks and about 50 men—were needed for each well and they were 'as fast, efficient, and cheap as the other more civilized methods now to be referred to'[83] The latter, two 40-cubic feet steam-worked Bell's dredgers, was 'only so much better than the primitive bullock system as to justify its retention as a reserve to overawe the bullock-drivers, who were a troublesome and bad-tempered lot, always ready to strike'.[84] The use of the presence of a more advanced technology to control the operators of a less-advanced and cheaper work process is fascinating. Spring clearly was a sophisticated manager who understood well the complex art of worker control. The two bridges were built expeditiously. Work on the Sher Shah began in March 1888 and the first goods train crossed the completed structure on 31 December 1889; the bridge across the Kistna was built between September 1890 and March 1893.

What shines through Spring's description is the extent to which bridge building in India as an organized process had advanced; how the construction process had become a better managed assembly of the various factors of production. The various elements and sequences of the process meshed smoothly even to the extent of ensuring a steady supply of stone coming into the bridge site at the rate of 1–3 trains daily, from quarries located miles away. It was even recognized that great care should be taken in Britain to design girders that would easily come together when erected in the conditions that prevailed at Indian sites.[85] The building of the Sher Shah and Bezwada bridges illustrates the systematic, thoughtful, continually refined application of the knowledge and techniques acquired over three decades of bridge building in India. The Bezwada bridge, in fact, added a future dimension to this systematization when the engineers considered certain aspects of its construction to be experiments designed to facilitate the building of the upcoming East Coast Railway bridge over the Godaveri (built 1897–1900).[86]

[83] Spring, Technical Paper no. 71, pp. 26–7.

[84] Ibid., p. 29.

[85] Cf. Berridge's criticism of Rendel's design of the Landsdowne bridge.

[86] See Technical Paper no. 26, F.J.E. Spring, 'The Design and Use of Caissons for

The centrality of better management rather than better tools to the successful construction of these two bridges was typified in the handling of the workers. The engineers, who, just as if they were contractors' agents, dealt directly with the workmen, were guided by the ruling philosophy: 'how shall I get it done best, most smartly, and cheapest, by direct dealings with the labourers'.[87] Spring paid attention to details that kept the work-force reasonably content and healthy. He found it expedient to build a flour mill at Bezwada to help keep his Punjabi workmen happy. They would not tolerate the rice diet of the locality and since the local wheat flour was 'wretchedly bad', Spring arranged for shipments of good wheat from Calcutta which was then milled at the bridge site.[88] In addition, careful attention to sanitation, water-supply, food and safety precautions exercised via 'a certain amount of paternal and disciplinary government' kept the work colony practically free of cholera and enteric diseases and kept serious accidents to six.[89]

Work processes were manipulated to the point where girder erectors were led to participate in the more rapid termination of their jobs. Spring cleverly observed:

> The way to erect economically is by any sort of inducement to get the several gangs of men to compete with one another, and this, is a thing which the natives of Northern India will always do if properly handled; the instant the work is completed the great mass of the labour can be paid off, and only a few selected men retained for the various finishing off jobs.[90]

The stages of construction of the two bridges replicate the activities at the building of most large bridges in India. First, there were the preliminaries: the establishment of brickfields, the collection of lime and stone, the building of houses for engineers and foremen and sheds for labourers, the construction of workshops, the assembly of stores and plant, and the laying of tramways to connect these various components with one another and with the site of the river crossing. Secondly, there was the construc-

---

Well-Foundations, as Adopted on the Construction of the Kistna and Godaveri Bridges' (1893); and Paper no. 86, F.T.G. Walton, 'The Construction of the Godavari Bridge at Rajahmundry on the East Coast Railway' (1900).

[87] Spring, Technical Paper no. 71, p. 17.

[88] Ibid., pp. 47–8.

[89] Ibid., p. 31.

[90] Ibid., p. 10. One can profitably read Spring's statement in the light of Burawoy's concept of 'making out' wherein 'game-playing' on the shop floor helps to manufacture consent among the workers to their own exploitation. See Michael Burawoy, *Manufacturing Consent* (Chicago: University of Chicago Press, 1979).

tion of the bridges' foundations. Finally, there was the erection of the iron work. Both bridges also involved river training works. The latter, Spring noted, benefitted from eighteen years of experience in India in the training of rivers during which engineers had learned to contain rivers within narrower channels and thus reduce the amount of expensive bridge work. This was the one area where cumulative advances in technical knowledge made an important contribution to the success of the two projects.

But, to repeat, it was in the management of the construction process itself that these two bridges best reflected the advance that had taken place. The process had become routinized; management had more effective control. The management, engineers and their subordinate supervisory staff, understood better how to put Indian labour to more productive use. Bridges in the 1890s were built more quickly than in the 1850s and 60s; labour was used efficiently and then dismissed.

The process of change came in evolutionary increments, but by the end of the period covered in this study, much had changed. Berkley and Spring were both competent engineers, but Spring had the benefit of some forty years of accumulated and collective experience on which to draw. Railway building had ceased to be an experimental art; it had become an applied science. But part of that science was the recognition of the need for flexibility. If Spring, at the Kistna bridge, exercised a close but subtle control, the engineers building the Assam–Bengal Railway's hill section (1896–1904) utilized Indian tools and appliances whenever possible 'and where they did not affect the progress of construction, native customs of labour or even prejudices were not interfered with'.[91] Marx was on the right track. Production, even in the inchoate setting of railway construction, had been partially transformed. But in nineteenth-century railway construction in India, at least, it had been transformed primarily through the more effective use of people which sometimes meant less rather than more control.[92] To paraphrase George Stephenson, engineers had learned to engineer men, and in India, women and children too.

Working Conditions

Railway construction was hard work. For the skilled and unskilled worker alike it was a tough, physically demanding job. Moreover, much of the work was inherently dangerous. Lack of care and precaution made it more

[91] Nolan, *MPICE*, 178 (1909), 322.

[92] From the perspective of capital the goal is not control of the labour process for the sake of control but control of the sort that leads to the desired result—in the case of railway construction, the expeditious and economical completion of a project.

so. The possibility of accidental death or injury was ever present. One surgeon reported that he had treated hundreds of BB&CIR workers with 'terribly severe injuries' received while employed on the Nerbudda bridge and adjoining works.[93] Bridge superstructures took workers to dizzying heights where little could be done to prevent the misstep that could send an individual plunging to certain death. Tools fell from the same heights and killed those below. Workers fell into rivers and drowned or water burst into foundation wells with the same result. Foundation work killed or crippled divers or workers in compressed air caissons. Careless or faulty blasting in tunnels or deep cuttings injured or killed others. Even without the dangers of blasting, tunnel work meant the possibility of cave-ins which could fall on the workers or trap them inside a shaft. Embankments collapsed, burying alive the unfortunate earthwork coolie, his wife or his child. Unguarded machinery could maim or snuff out the life of the unwary. The careless use of construction locomotives or inattention on the part of the victim, also took the life or limbs of others.

We do not know how many people died or were injured in these ways. Extensive data on accidents of all sorts on the operating railways came to be collected and published regularly by the government, but the same effort was not made with respect to construction workers. In an isolated example which reveals something of the extent of injuries and deaths, it took hundreds of pages of material in 1859 to describe construction accidents on the GIPR, the BB&CIR and the Sind line of the SP&DR.[94] Otherwise we get scattered reports containing actual numbers or references to considerable loss of life.[95] Six were reported to have died an accidental death on the Lansdowne bridge; nineteen died on the Dufferin bridge but that was explained by the fact that 'a far larger number of hands was employed than would be the case on such works in Europe . . .';[96] 4857 accidental injuries occurred during the construction of the meter gauge West of India Portuguese Railway and its accompanying harbour works.[97] Also revealing of the often dangerous conditions of railway

[93] MSA, PWD (Railway) 1859, vol. 44, compilation 309: 'Remuneration to Drs. Dent and Pinkerton for their professional services to the BBCIR'.

[94] MSA, PWD (Railway), vol. 20, compilations 88, 89, 224, 361, and 384.

[95] The latter example is from the Bhore Ghat. Berkley, *MPICE*, 19 (1859–60), p. 608.

[96] Walton, *MPICE*, 101 (1890), p. 24.

[97] Sawyer, *MPICE*, 97 (1889), p. 322. In total the medical staff treated 63,000 patients of whom 392 died during the four-year project. Fever was most common, and accounted for 25,144 patients. A far larger number attended to themselves in their huts and did not become a part of these statistics.

building is the correspondence between the Chief Engineer of the IVSR and the government over the need for a suitable medical establishment near the proposed line of works near Sehwan; 6000–8000 work-people were expected on the dangerous (heavy cuttings through rock) section once work was in full progress to whom accidents would 'inevitably happen' creating the need for 'a special surgical hospital'.[98] Perhaps, however, the safety record did improve with time. Few died at the Bezwada bridge and only three men lost their lives—two in a suddenly flooded well—and only one was seriously injured, a lost leg, during the construction of the new Chakdara bridge over the Swat river, north of Malakand, in 1901–3.[99]

Usually records only provide scattered, grim totals whose cold numbers sanitize and conceal the death and suffering they represent. We rarely know the specifics of an accident and even more rarely do we know the identity of the victim(s), especially Indians. The injury of three coolies and the death of two women and a man in an accident on the Thal Ghat in 1859, when an embankment collapsed, is about as detailed a record as we get.[100] Accidents to Europeans were reported somewhat more fully. We know that D. Syme, a Scottish fitter attending to an engine on a boat used in well-sinking operations for the PNSR's Chenab bridge, died on 30 March 1874, at the age of 28 years and one day. He fell overboard and, being unable to swim, drowned before he could be helped.[101] We can even speculate that Mr Syme fell because he was still enduring the consequences of a robust birthday celebration the night before. But can the historian express the terror Mr Syme must have felt? Can the historian reconstruct the agony of a diver dying from the bends under the medical conditions present in rural India in the 1880s? The extended moment of horror of the iron-worker falling from a high girder during the erection of the Lansdowne bridge? Or the pain of coolies suffocating under the earth of a caved-in embankment? I cannot, but I do know that we, readers and author alike, must not forget that building India's railways took a terrible toll about which there was nothing abstract: individual men, women and children paid the highest price for those railways.

[98] IOL&R, P/558, PWD, RR Estabs. Progs., June 1874, no. 15.

[99] 'Construction of the Chakdara Bridge', Professional Papers of the Corps of Royal Engineers, Royal Engineers Institute. *Occasional Papers,* vol. 29 (1903), p. 122.

[100] MSA, PWD (Railway), 1859, vol. 20, compilation no. 224: 'Accidents on the Lines of Railways under Construction'.

[101] IOL&R, P/558, PWD, RR Estabs. Progs., June 1874, nos. 1–3.

Accidents, however, were not the main killers and cripplers; diseases were. Cholera, malaria, small-pox, typhoid, pneumonia and many other diseases caused an enormous loss of life. Others were stricken and eventually recovered or, particularly in the case of malaria, were doomed to a shortened, debilitated life. Cholera was the great killer that came in epidemic gusts and caused coolie gangs to scatter in the hope of escaping the deadly disease. Thirty per cent or more of the workers on a section of line might die within a few weeks. One can pile example upon example of the effects of cholera on construction gangs but one extended account best captures the horror of a cholera epidemic, compounded, in this example, by other diseases.[102] This occurred during the construction of the Bengal Nagpur Railway in the Umaria District in 1888:

All the year round the whole country is very feverish, the months of January and February being slightly less so. The 'Lanias' who come down from 'Purtabgarh', 'Jownpur' and other adjacent districts in Oudh die in great numbers from fever, dysentery and ulcers. These latter are caused by the slightest abrasion of the skin from a flying stone or other slight cause, and in a few days whatever the reason may be, the part becomes a foul ulcer which unless treated at once take months to cure, if indeed they do not prove incurable. Men may be seen with their legs or arms almost rotting off with these sores. Between the months of March and July the water-supply fails almost entirely for miles, the streams either dry up or become so foul from rotting vegetation that the water is undrinkable. The rainfall of 1888 was much less than the average and the water-supply failed in April, hence in May cholera broke out at 'Pali' and spread over the whole length of the district, the mortality being terrible, some 2,500 or 3,000 labourers and petty contractors being carried off by it; the scare was such that people hurrying away at the first sign of its approach were left by their relatives to die all along the line, giving the staff (the Assistant Engineers having in dozens of instances to drag the bodies away themselves, make the funeral piles and light them with their own hands) as much as they could do and more to keep the road clear. None of the natives, with but the rarest exceptions, would touch the corpses or go within fifty

[102] For examples in Bengal, 1853, Bhore Ghat, Thal Ghat and other GIPR contracts, 1859–61, EIR, Jubbulpore Extension, 1863, SIR, Madras–Cuddalore line, 1875, Bellary–Kistna State Railway near Guntakal, 1886, Sind–Peshan State Railway, Bostan–Gulistan section, 1886, and the SIR, Pamban branch, 1900 see respectively: *Friend of India*, 26 May 1853, p. 321; MSA, PWD (Railway) 1859, vol 44, compilation 441, 1860, vol. 30, compilations 354 and 401, and 1861, vol. 7, compilation 456; IOL&R, P/191/4, PWD, RR Progs., February 1864, no. 7; IOL&R, L/PWD/3/219, Madras RR Letters, no. 7 of 1876 dated 16 June 1876; IOL&R, PWD, RR Construction Progs., February 1886, no. 194; IOL&R, PWD, RR Construction Progs., August 1886, no. 297; TNSA, Madras PWD (Railways) 8 January 1901, nos. 14–15.

yards of them. (It may be mentioned as a curious fact that during the outbreak of cholera no vultures, the usual useful scavengers, were to be seen either at their work or on the wing.) In one gang of 30 men employed breaking ballast 14 were carried off in two days, also Sub-Overseer. At the Gorchetta Bridge on which the girder erection was going on at the time some 50 men were carried off, the contractor himself was taken ill, but ultimately recoved [sic]. The whole line was deserted throughout the district. Mr Thomason, Assistant Engineer, after giving medicine &c., to a number of his men who were down at Anukpur, where the foundations were being taken for a bridge, himself succumbed in about six hours before any of his brother officers could reach him. Even now, nearly three months after the cholera has practically ceased, on pulling down the little grass shelters erected by the coolies, numbers of skeletons have been found. The greatest difficulty was experienced in procuring provisions as all the petty villages within 5 or six miles of the line were closed to outsiders, the villagers declining to allow even a European inside, and ready even to use their axes and latties to prevent it. One village or rather hovel containing originally eighty people all told, now contains 28, the rest are dead.[103]

The techniques of railway construction, the patterns of worker recruitment and the living conditions at the work-sites increased the possibility of outbreaks of certain kinds of diseases and then further increased the possibility that once an outbreak occurred, it would spread quickly among the assembled workers and beyond to the local populations. The construction techniques provided favourable conditions for the breeding of malaria-carrying mosquitoes. Earth for railway embankments often was dug from borrow pits along the line-of-works. These pits filled up with water and vegetation, and became mosquito hatcheries, as did the waters impounded by the embankments whose inexorable progress across the countryside so often interfered with the natural lines of drainage, created ponds and raised sub-soil water-levels.[104]

The patterns of worker recruitment also contributed to the spread of disease. Many railway construction workers were highly mobile; they brought diseases to the work-sites and carried newly acquired diseases onwards to other sites.[105] Locally recruited workers returned to their

[103] IOL&R, P/3469, PWD, RR Construction Progs., October 1889, no. 490.

[104] A fine summary of the effects of railway construction and other public works on the spread of malaria can be found in Sir Patrick Hehir, *Malaria in India* (London: OUP, 1927), pp. 45–9. The situation he described in the 1920s was 'better' than that present in nineteenth-century railway construction when less was known about the cause and prevention of malaria and other diseases.

[105] The vulnerability of the construction workers to disease was shared by other migratory workers. Ralph Shlomowitz and Lance Brennan, 'Mortality and Migrant

villages on a daily or weekly basis and carried disease germs back and forth between the work-sites and villages. Large concentrations of workers also required provisioning arrangements which, whether planned or unplanned, meant the appearance of a nearby bazaar where, again, transmission of diseases among different bodies of workers, merchants and local residents took place.

Given the endemic presence in nineteenth-century India of a number of potentially epidemic diseases, the construction techniques and patterns of worker recruitment increased the possibilities of epidemic outbreaks. The living conditions at the work-sites brought this dangerous combination to its explosive culmination in the repeated outbreaks of cholera, malaria and pneumonia that killed many railway construction workers and others who came in contact with them. Life at the work-sites was grim and hard for the poverty-stricken, malnourished, weakened, disease-susceptible men, women and child labourers. What the epidemiologists came to call 'large aggregations of tropical labour' lived in crowded, unsanitary conditions that virtually guaranteed a rapid spread of diseases.[106]

Not uncommonly, 75 per cent of the inhabitants of the Adamwahan construction colony, adjacent to the Empress bridge (built across the Sutlej, 1874–8), by no means the least regulated of such sites, simultaneously were incapacitated by fever in the worst part of the working season, while in 'one year, when a flood had broken into the place, one thousand labourers are believed to have died of pneumonia'.[107] The Bhore Ghat, with its immense concentrations of labour, was the site of repeated outbreaks of cholera, though this did not surprise a government consulting engineer who reported in May 1860: 'Everyone who has visited the works must have been sensible to the want of the most ordinary sanitary

---

Labour en route to Assam, 1863–1924', *IESHR*, 27:3 (July-September 1990), pp. 313–30, provide detailed statements for the tea-plantation labour. Again, cholera was the great killer.

[106] Hehir, p. 46 has an excellent passage which, in so far as malaria is concerned, effectively and graphically describes the 'accumulation of circumstances' in the labour camps that was largely responsible for the virulence of the infection in the wet season and the maintenance of a reservoir of residual malaria in the dry season from which new outbreaks occurred in the next malaria season.

[107] Bell, *MPICE*, 65 (1881), p. 257. Note that a 1000 people were 'believed to have died'. Perhaps the number was larger or smaller, but the Europeans did not know for sure. 'Fever' was a residual category for all sorts of illnesses but often referred to malaria.

…ents necessary where large bodies of natives are encamped long …pot. The offensive smell arising from the nuisances committed by thousands of men crowded in a small space, always suggested the probability of epidemic disease.'[108] But despite reasonable sanitary arrangements which kept cholera in check on the Chaman Extension (1887–91) of the Indus Valley line, at least 800 died in the winter of 1890–1 from typhus fever brought to the camp by workmen from Kandahar.[109]

Malaria, more insidious and less understood, took its share of lives as well. A twentieth-century malaria expert suggested that 'a death a sleeper' (and some 1700 sleepers were needed for each mile of track) was a vivid, if difficult to substantiate, estimate of the toll malaria exacted on the Ghat sections of the GIPR.[110] The same authority reported that when the Saranda Tunnel carrying the Bombay–Calcutta main-line through the Saranda range in the Singbhum district in Chota Nagpur was constructed in the 1880s 'mortality and bolting among the labour so disorganised the accounts that the gangs were paid, shift by shift, as they emerged from the workings. Several of the Cornish miners who formed the subordinate tunneling staff died of malaria, or more probably of blackwater, for which the district is notorious.'[111]

Most of the conditions just described were present during the construction (1856–63) of the fifteen miles of the Bhore Ghat incline that carried the south-eastern line of the GIPR up the precipitous Western Ghats. To work on the Bhore Ghat incline was to work in a crucible of suffering. Work in the hazardous and isolated conditions of the Ghat was extremely dangerous. On some cliff faces no footholds existed, so workers had to be suspended by ropes in order to drill and blast the right of way into the rock. Sometimes, a contractor's agent reported, a worker 'would loose his hold and get dashed in pieces in the nullahs below', which 'had the effect of deterring his fellows, altogether, from working for days'.[112] The hard rock

[108] MSA, PWD (Railway), 1860, vol. 26, compilation 332: 'Cholera at the Bhor Ghat'.

[109] Technical Papers no. 35: 'Report on the Chaman Extension Railway' by G.P. Rose (May 1894), p. 3.

[110] IOL&R, V/25/720/8, Railway Board Technical Paper, no. 258: R.S. White, *Studies in Malaria as it Affects Indian Railways*, part I (Calcutta: Government of India, Central Publications Branch, 1928), pp. 1–2. White was the first malaria specialist to be appointed by an Indian railway—the Bengal–Nagpur line.

[111] Ibid., p. 20. Blackwater fever or *kala azar* is caused by a distant relative of the malaria parasite and is probably transmitted by a kind of sand-fly.

[112] MSA, PWD (Railways), 1859, vol. 25, compilation 206: 'Bhor Ghat Incline. Surrender by Mr. Faviell of His Contract for the Construction of the Work on', E. Swan to Faviell dated 3 November 1858.

demanded the extensive use of blasting powder which resulted in a considerable loss of life.[113] Falling rocks, slips and cave-ins were continually a problem and a threat to lives.

Living conditions were primitive, difficult and often deadly. Despite the heavy precipitation during the rainy season, many parts of the Ghat were waterless in the dry main working season, thus requiring supplies of drinking water to be carried laboriously to the assembled workpeople. Strong winds buffeted the workers and their flimsy, lean-to dwellings. As described above, sanitation was almost non-existent and disease took a dreadful toll. The frequency of accidental death and injury was a small number when compared to the mortality and morbidity rates caused by cholera and other diseases. Cholera ravaged the workforce almost from the inception of the project, but the intensity of the epidemics mounted in the later years as the number of labourers crowded upon the works increased. Ten workers per day died during an outbreak in late 1859 which continued into January 1860 and flared up again in April–May 1860.[114] The latter outbreak killed 25 per cent of the Europeans and 'of the natives so numerous as to be beyond accurate calculation'.[115] A milder outbreak occurred in the 1860–1 working season. The distance of the Ghat works from public thoroughfares and the workers' practice of living in little colonies near their respective works which they never left until the end of the working season meant that the authorities were slow to hear of, and thus slow to respond to, outbreaks of disease.[116] Each major epidemic caused the workers to flee, thus slowing or even stopping construction until the labourers could be induced back to the incline. 'Jungle fever' (possibly malaria) took its toll as well—and when it did not kill people, it disabled them. A contractor's agent reported in 1858 that during the eight month working season on the Ghat (the rains halted worked each year for some four months except in the tunnels where it went on year round), European agents were, on an average, disabled by sickness for some six weeks.[117]

Conditions on the Ghat bred violence and lawlessness. People on the

[113] Berkley, *MPICE,* 19 (1859–60), p. 608.

[114] MSA, PWD (Railway) 1859, vol 44, compilation 441; 1860, vol. 55, compilation 224; and 1860, vol. 26, compilation 332.

[115] MSA, PWD (Railway) 1862, vol 6, compilation 317. This information certainly supports the possibility that upwards of 25,000 lives were lost during the construction of the incline.

[116] MSA, PWD (Railway) 1860, vol. 26, compilation 332.

[117] MSA, PWD (Railway) 1859, vol. 25, compilation 206, Appleby to Fowler dated 31 October 1858.

margins of Indian society—tribals and members of low or untouchable castes—were supervised by a rough and ready lot of Europeans, within a harsh human and physical environment. A British official observed, after violence led to the death of a European, that it was a wonder 'considering the large mass of labourers brought together . . . on the Ghaut . . . almost totally removed from public observation, and shamefully treated as there can be no doubt those labourers have been' that an outbreak had not come sooner.[118]

Relationships between the few permanent inhabitants of the Ghat and the seasonal construction workers were also strained. Workers were implicated in robberies and various expedients were suggested for their better control, including appointing a mucaddum for every fifty workers to ensure that they were in their hutments after 8.00 p.m.[119] 'Notorious lawless tribes' like Ramoshis and Kaikaris were to have special muccadums appointed by the police.[120] Some Ramoshis, in turn, acted as watchmen on the Ghat works.[121] The construction of the Bhore Ghat incline provided the experience upon which the British authorities formed the legal and police structure that was to govern behaviour at other work-sites and regulate relationships between employers and employees. Violence, oppression and lawlessness did not disappear from subsequent work-sites, but they were less evident and less frequent. This was partly because of the lessons learned and the arrangements put into place during the course of the work on the Bhore and Thal Ghats, which helped to ameliorate the worst excesses; it was also partly because the particular combination of massed labour and deplorable conditions present on the Ghats, were never

[118] MSA, PWD (Railway) 1859, vol. 25, compilation 215: 'Bhore Ghat. Disturbances amongst the workers'.

[119] MSA, PWD (Railway) 1862, vol. 36, compilation 227: 'Robberies in the vicinity of the Railway works on the Ghats'.

[120] The presence of Ramoshis among the workforce is interesting. They were one of those groups the colonial state officially designated a 'criminal tribe'. They were particularly located in Poona, Satara and Ahmednagar, i.e., not too distant from the Ghat works, but they were not particularly associated with earthworking though some were field labourers. In Maratha times they were in charge of hill forts. See R.E. Enthoven, *The Tribes and Castes of Bombay* (1922, reprint edn, Delhi: Cosmo Publications, 1975), pp. 297–304. Enthoven states that their hereditary occupation was stealing—an example of the ethnographic stereotyping that went on in colonial India. The making of the 'criminal tribes and castes' in the colonial discourse is increasingly subject of a study. See, for example, the two part study by Sanjay Nigam, 'Disciplining and Policing the "Criminals by Birth" ', *IESHR*, 27:2&3 (April-September 1990).

[121] ICE, *Paper on the Bhor Ghaut*, p. 47.

so intensely present in future construction. The workers on the Bhore Ghat were intractable throughout the construction process and with good reason.[122]

The descriptions above attest to the fact that early in India's railway age the British understood something of the conditions that led to epidemic diseases among the construction workers. The etiology of cholera and other diseases spread primarily through the intestinal tract was still unknown, but there was a recognition of the fact that unsanitary living conditions had much to do with the problem. The role of the mosquito in the transmission of malaria was not discovered until 1898, so well-informed, preventive measures against this scourge were not undertaken in the nineteenth century. However, the prophylactic effect of quinine against malaria was known, but it was not used on a wide scale in the nineteenth century to protect Indian labourers or, in the early going, European railway engineers. Regardless of the many complex reasons that retarded the advancement of public health in nineteenth-century India, those in charge of building India's railways slowly felt a greater concern for the physical well-being of construction workers.[123] Medical personnel became a part of the establishments approved for lines under construction. Health rules were published that the engineering staff were supposed to enforce with respect to the location and conduct of bazaars, the provision of drinking water, and the placement and construction of latrines.[124] A concern for sanitation and conservancy then found its way into the lectures and manuals prepared for the instruction of the young engineer and into practice at the work-sites.[125]

The main motive for this accumulating concern for the health of the

[122] In 1858 Appleby complained about 'acts of insubordination' and the frequent insults Europeans had to bear. A serious affray at Khandalla in January 1863, during which some Europeans were beaten and which had as its background bad blood between the villagers and the railway workers (both Indian and European), testifies to the continuation of unrest. See, respectively, MSA, PWD (Railway) 1859, vol. 25, compilation 206, Appleby to Fowler dated 31 October 1858, and *Times of India*, 4 March 1863, p. 2.

[123] David Arnold, 'Medical Priorities and Practice in Nineteenth-Century British India, *South Asia Research*, 5:2 (November 1985), pp. 167–83, discusses some of the reasons for poor public health.

[124] For example, IOL&R, P/2751, PWD, RR Construction Progs., September 1886.

[125] IOL&R, 300 A. 78.F. (c). Sir J.R. Wynne, *Notes on Construction of Railways in India*, (Calcutta, 1902). Better conservancy and its positive effects can be followed in Martin-Leake, *MPICE*, 151 (1903), p. 277 and Gales, *MPICE*, 174 (1908), p. 37.

worker was the recognition that the prevention of epidemic disease was an important element in obtaining and retaining labour at the work-sites and thus in getting the job finished on time; that 'economy of execution' and 'the health of labour forces' were linked.[126] As Addis put it: 'The Indian workman is easily scared—and who can blame him—and at the first sign of cholera crowded areas are depleted, practically instantly. . . .'[127] The results of the increased concern for the health of the construction workers, however, were limited in the nineteenth century. Throughout the period covered in this study, Indians were extremely vulnerable to epidemic diseases which frequently struck the railway work-sites and other large-scale public works. The fundamental reason is not hard to discover. As Hehir observed in the 1920s: 'Here we have large numbers of people, poverty-stricken and with lowered physiological resistance, working hard. . . .'[128]

The construction gangs lived a dangerous and unhealthy life: disease was their constant companion; death by sickness or accident frequently visited them. The gangs also lived a life which was punctuated by violence and characterized by oppression. The violence came in many forms. The subordinate European supervisory staff, the foremen and such like, sometimes assaulted Indian workmen. They were particularly prone to do so when drunk. One magistrate reported, after visiting Khandalla on the Bhore Ghat, that Indians had to be better protected 'against the license of Europeans employed on the Railway Works', and that 'several Europeans of a secondary rank in life, with habits and means of drinking, cannot safely be left among a Native population unwatched by Magisterial authority'.[129] Other assaults, however, can only be explained by the trying conditions at work-sites and by racial prejudice, compounded by a Briton's frustration at his inability to communicate and command across the barriers of language and other cultural differences that separated him from the Indians he supervised.[130] Moreover, Britons who occupied lower level positions came from the British lower classes whose

[126] Evans, *MPICE*, 200 (1915), p. 2.

[127] Addis, p. 119.

[128] Hehir, p. 46.

[129] MSA, PWD (Railways) 1858, vol. 18, compilation 294: 'Complaints. Misconduct of Certain Europeans in the Service of the Railway Contractor at Khandalla'.

[130] Since most Europeans did not physically assault workmen—though they may well have shared a racially inspired dislike of Indians with those who did—we must conclude that idiosyncratic factors beyond historical reconstruction were also at work. Put more simply, some people are more ill-tempered than others.

values were more likely to countenance personal violence as a solution to difficulties—a tendency enhanced in a situation where the 'cause' of one's frustrations was perceived to be inferior.[131] This kind of explanation would seem to fit best the example of violent behaviour by a Mr Sgardelli, an inspector of the Hooghly Bridge Works, who had hitherto 'borne a good character' and other examples of Europeans assaulting Indian workmen where drink was not involved.[132]

On 6 June 1884, Sgardelli assaulted a Bihari named Sheik Sobhan who was carrying concrete from a boat to a pontoon attached to the bridge. The two argued, the inspector hit Sobhan on the chest, shoulder and head, and then kicked him. Sobhan fell into the water. A body was subsequently found and the medical examiner testified that the man had not died from drowning but was dead before falling into the river. But was it Sobhan? Another khalasi had been struck by a bridge crane and thrown into the Hooghly on the same day. A second body was found later, but both bodies being much disfigured, identification was uncertain. Sgardelli was tried and acquitted by a jury of the three charges against him: culpable homicide not amounting to murder, grievous hurt and simple hurt. The judge agreed with the first two decisions but not the third—simple hurt—but had no option other than to accept the jury's verdict. The EIR subsequently tried and failed to get the government's permission to write off one-half of Sgardelli's substantial legal costs against the expenditure on the bridge works. The government consulting engineer, in forwarding and supporting the request to government, noted that there was no doubt that Sgardelli had struck Sobhan and that he deserved to be punished departmentally for the offence.

Violence sometimes begat violence as workers responded in kind to European assaults. More common was the violence of worker against worker or the violence of migrant workers against neighbouring resident populations and vice versa.[133] Sometimes, too, local criminal elements

[131] I do not mean to suggest that these lower class Britons were intrinsically more violent than their higher class brethren. Their life situation made them more prone to use personal violence. The upper classes had other, more effective, methods of getting their way.

[132] IOL&R, P/2536, PWD, RR Estabs. Progs., January 1885, nos. 1–4. There was one other re-occurring cause of such attacks. Europeans sometimes had a mental breakdown; they became 'mentally deranged' in the vocabulary of the period. See, for example, the case of A.F. Trench, an assistant engineer engaged on irrigation works. IOL&R, P/558, PWD Progs., October 1873, nos. 4–6.

[133] The relations in production certainly include worker to worker relationships.

preyed on the workers' encampments. What must be kept in mind was that a major project crowded large numbers of people—mostly strangers—together in isolated areas and under stressful and deplorable conditions. These people were, in turn, for the most part from the bottom, impoverished levels of Indian society. It is perhaps surprising that railway construction in India was not characterized by more violence than in fact took place. Certainly, the navvies who built canals and railroads in nineteenth-century Britain, Canada and the United States, were a more violent lot in many respects.[134]

Oppression and exploitation was more subtle and pervasive. At one level they were a feature of the wage relationship between the poor worker and the person who employed him. The worker had to sell his labour power to employers who took little if any responsibility for the worker's well-being and who, hiring on a short-term basis, dismissed the worker or the gang of workers once it was monetarily advantageous to do so. The worker usually needed the job more than the employer needed a particular worker. The worker might also be indebted to the sirdar or muccadum who had advanced the money that enabled the worker to get to the work-site in the first place. Wages were low, often in arrears, and sometimes paid short. The gangers took their 'share' of the workers' wages. The truck system represented another form of exploitation.

This chapter has described something of the work of construction itself and the conditions within which it was carried out. The work has been explored primarily from the perspective of the management of the work processes, while the brief description of the conditions of work has stressed the physical and human causes of the hard and dangerous lives led at the work-sites. Both aspects, however, suggest another likely dimension of the work situation: worker resistance to changes in the labour processes due to unhappiness over low pay or wages in arrears, or to fear stemming from the conditions of life and work at the construction sites. This dimension, surely, must have affected the ability of the British to build Indian railways expeditiously and/or economically. It is to worker resistance, therefore, that we shall briefly turn, in the next chapter.

[134] An example is to be found in Peter Way, 'Shovel and Shamrock: Irish Workers and Labour Violence in the Digging of the Chesapeake and Ohio Canal, *Labour History,* vol. 30:4 (Fall 1989), pp. 489–517. Way's footnotes lead one into the literature on canal construction labour in Canada and the U.S.A.. Coleman, *Railway Navvies,* provides examples of violence among the navvies in British railway construction.

CHAPTER 6

Worker Resistance

The Indians who constructed the railways did not passively accept all the dictates and nuances of the new or modified labour processes in which they were engaged—the oppression, exploitation, brutality and harsh conditions at work-sites. Nor did they accept, within the limits of what they could do, the epidemics of disease; they fled from work-sites and returned, if at all, slowly and reluctantly, while fresh workers proved most unwilling to go to areas known to be especially unhealthy. They did on many an occasion act to right an injustice or improve their material position. Although we can only know of these resistances as they come to us, refracted through British eyes and minds, from English-language sources, their presence is clear. These episodes certainly did not represent the workers' conscious resistance to capitalism, but given the subjective state of the workers, their previous history, and the general and particular conditions, both social and technical, under which they worked (including the ways in which the work was organized and supervised), we can be certain that we are examining manifestations of the fundamental opposition of interest between 'those for whose purposes the labour process is carried on, and those who, on the other side, carry it on'.[1]

Given the perspective taken in this book, this chapter does not focus on the sources and forms of worker agency located, *inter alia*, in the normative assertions of a moral economy, the power of custom and established behaviour, and, more generally, in the inherited and on-going weight of cultural practice and belief that, for some, is the essence of structuralism.[2] A study of worker resistance from that perspective would take us—and may take a future historian—into the subaltern world already opened up by others.[3] This book focuses on how the British got

[1] Harry Braverman, *Labor and Monopoly Capitalism* (New York: Monthly Review Press, 1974), p. 57.

[2] Marshall Sahlins, *Culture and Practical Reason* (Chicago: University of Chicago Press, 1976), especially ch. 1.

[3] Notably the subaltern studies group whose investigations have appeared in a

the railways of India built, and from that point of view it is sufficient, firstly, to establish that among the obstacles the British faced was worker resistance of various sorts; secondly, to categorize the types of resistance that occurred; and, thirdly, to show how the British dealt with worker resistance.

Resistance appeared in many forms, from what could almost be characterized as 'inaction', to direct, sometimes violent, action; from subtle and complex attempts to control a work process, to immediate forceful responses to a brutal supervisor. In concrete terms, some potential workers refused or tried to refuse construction work; some workers joined and then exited; others stayed and gave voice to protest through a variety of methods ranging from petitions and peaceful refusal to accept a new tool, to the more vehement voice of the strike or physical confrontation.[4] The causes which triggered resistance were also diverse and ranged from those diffusely present in the general social conditions to specific ones of wage rates and the demand for regular wage payments.

One can argue, however, that the more specifically the cause of resistance related to the control of work and to the size of wage payments for that work, the more the workers can be seen to have been inducted into the social relationships of capitalism. The circulating workers (for whom exiting or never starting was not a good option) who focused on these issues had more fully accepted their emerging proletarian position. Clearly, the workers subjective and objective conditions, prior to and during railway work, was an important determinant of their behaviour. For example, some poor peasants and landless labourers could not or would not become full-time construction workers because their bondage to their village was too strong to make the break—a bondage that could be objective, e.g. landlord power, or subjective, e.g. the belief that the village still offered a more secure material existence (or since human existence is rarely so tidy, both). Some in fact worked on the railways temporarily and involuntarily because village power-holders made them do so. A surveyor of the proposed Bombay to Neemuch line expected that the earthwork would be done cheaply if the Thakurs were persuaded to

---

series of volumes edited by Ranajit Guha, *Subaltern Studies. Writings on South Asian History and Society,* vols I–VI (Delhi: OUP, 1982–9).

[4] Although his point is vastly different from mine (or Braverman's), Albert O. Hirschman, in his *Exit, Voice and Loyalty* (Cambridge: CUP, 1970) provides a useful way of categorizing the different forms of protest I adapt. The book deals with how and why people leave, protest or remain loyal—as customers for goods and services, as citizens of states, or as members of organizations.

do the work by contract through their own districts, because 'there is not a Takoor in this part of the country that cannot command from 500 to 1000 Bheels all of whom go to work for him at a moment's notice receiving no other remuneration for their labour than one meal a day'.[5] Others, including those who were already migrant workers, as this book has demonstrated, became committed (not in a psychological sense but in the sense of economic necessity) to the life of the construction worker, while yet others became circulating workers through the impact of the railway construction labour market. The surveyor just quoted—and this is in 1865—reported that masons along the proposed line were 'becoming a migratory class of men', and their labour could be procured without difficulty.[6]

The first form of resistance, therefore, was to the process of proletarianization itself; it was resistance in the form of refusal to work at all (not refusal to work without higher pay) or refusal to work beyond a certain amount, and departure (exit) from the work-site if more was asked. This was either the resistance of those who were not keen to work in railway construction and did not have to do so; or, since railways were a leading edge of the penetration of advanced capitalism into the vitals of rural India and thus into the gut of an existing mode of production, resistance from those in local command of the existing mode by way of limiting the participation of their inferiors in railway construction.[7] The latter kind of resistance, however, could be overcome by the co-option of village power-holders. As one engineer astutely observed—as did the surveyor of the Neemuch line mentioned above—the lack of labour for railway construction in the Nizam of Hyderabad's territory could be overcome if the zamindars and headmen came forward as contractors: they would ensure that labour was procured.[8] Such a solution met the needs of everyone except the labourers themselves: the railways got their construction labour and landlords retained their hold over their artisans, tenants, and landless labourers.[9] These forms of resistance appear to have been more

[5] IOL&R, L/PWD/3/273, Bombay RR Letters, Enclosures to letter 10 of 1865.

[6] Ibid.

[7] Frankly, I do not care how one labels that existing mode, and thus dismiss a tedious part of what is, in some other respects, the important mode-of-production debate. Whatever the existing mode was, it was not industrial capitalism.

[8] IOL&R, L/PWD/3/218, Madras RR Letters, Collection to Letter no. 2 dated 13 February 1864. There is also an interesting example from Guntur district.

[9] Indian landlords were not the only ones to take up small railway contracts to ensure their hold over their labourers. European tea planters in Assam did the same

common in the early decades of the railway age. They demonstrate how provisional, tentative and uneven the advance of capitalism was in nineteenth-century India. I do not take the absence of documented cases of such resistance later in the century as an indication of the fact that the transition in many rural localities had gone much farther. It was more an indication of the extent to which railway building had come to depend less on locally recruited labour.

Resistance to proletarianization was seen in the situation described by Charles Henfrey (see Chapter four) on the Delhi–Amritsar contract where, he claimed, it was often necessary to drive the coolies out of their villages in the morning to force them 'to earn a good day's wages on the neighbouring railway works'.[10] This also reveals one forceful response to the resistance. Henfrey attributed the resistance to the villager's preference for 'perfect idleness', but we can see it, more correctly one suspects, as resistance to an unpleasant and alien form of work. The same complaint was made by engineers supervising the construction of the GSIR near Tanjore in 1860. People were 'indolent and independent' and would work only for high pay and only under the condition that they 'be allowed to leave and resume work at such times as suited them'.[11] A similar kind of complaint was made at about the same time by engineers in Sind who claimed Sindhi labourers were not to be depended upon and that they left if extra work or longer hours were demanded from them.[12] The same complaint was still being made in the 1890s by engineers in the Assam Hills.[13]

For those who had other options, railway construction work was not

---

thing in the 1890s. See IOL&R, P/4586, PWD, RR Construction, July 1894, no. 292.

[10] IOL&R, Tracts, vol. 592: 'Opening of the Meerut and Umballa Section of the Delhi Railway, On the 14th of November 1868' (London: W.H. Allen & Co., 1869), p. 32.

[11] IOL&R, L/PWD/3/217, Madras RR Letters, Collection to Letter no. 8 dated 30 April 1860. Such a condition strikes at the essentials of the relationship between capital and labour. How can the capitalist realize the potential of the labour power he has purchased when the worker insists on such control of the labour process. Of course, the 'indolent Indian' was a colonial stereotype that began well before the railway age. The stereotype is explored in Syed Hussein Alatas, *The Myth of the Lazy Native* (London: Frank Cass, 1977).

[12] IOL&R, PWD, RR Progs., 19 April 1861, no. 73.

[13] Local labour refused to begin work before 8.30 a.m. and quit working altogether when the work moved five miles beyond their village. IOL&R, P/5003, PWD, RR Construction, July 1896, no. 252.

necessarily attractive. Such people were semi-proletarians fc railway work was a temporary and supplementary activity; they remained more involved in the economy of their village. They, or those who commanded their labour, were not adverse to withholding their labour from engineers and contractors in situations where they believed the railways desperately needed them—even semi-proletarians saw the value of higher wages and understood the leverage that a high demand for labour in a particular area provided. Early in the 1860s, coolies in the Trichinopoly district of Madras and those along the line-of-works of the Eastern Bengal Railway were able to combine sufficiently and stay away *en masse* from the railway works, in the hope of raising earthwork rates.[14] Their ability to do so, of course, depended on the fact that they did not need the railway work to subsist. And, though it is not mentioned in these particular cases, it is possible that it was petty contractors who were behind the attempt, since widespread combination away from the work-sites suggests an organizing agency. In some instances, contractors were known to combine in an effort to force the railways to concede to higher rates.[15]

However, as argued in Chapter four, the railways came to depend more upon circulating, 'free' labour for their construction needs. These workers more fully accepted their role as waged labourers, and the skilled workers among them came to have a considerable stake in the railways' employment of their particular skills. Acceptance of their role as waged labourers, however, did not mean a lack of resistance to the conditions of work.

One continuing form of resistance, so unobtrusive as to be easily overlooked, deserves examination before the more dramatic manifestations of collective action are described. This was the resistance to certain specific innovations the engineers wished to introduce, i.e. a new tool, a new way of working or a new form of payment for work. Some of this resistance represented the disinclination of workmen to alter long-standing patterns with which they were closely familiar and comfortable. This kind of resistance, CE Brunton believed, could usually be overcome. He stated:

[14] IOL&R, L/PWD/3/217, Madras RR Letters, Collection to Letter no. 8 dated 30 April 1860: 'The work was delayed for nearly six months by a strike and a combination among the coolies for the purposes of raising the prices of earth work.' Also IOL&R, L/PWD/3/62, Bengal RR Letters, 1863: 'They [villagers] are at all times very shy and suspicious of strangers, besides which they were waiting to see what prices could be obtained by holding back.'

[15] See, for example, IOL&R, P/2540, PWD, RR Accounts, December 1885, no. 72 and P/2751, PWD, RR Construction, September 1886, no. 232. In the first example, from the Narainganj–Dacca–Mymensingh State RR, a less than expected

The Cutch carpenters and smiths, as has been before remarked, are intelligent and excellent workmen; but they were wedded to the use of their own rude tools, in their own fashion, involving great delay in turning out work. Their usual method is to carry on all operations while seated on the ground; and, in the case of the carpenters, to make almost as much use of their toes as of their fingers. It became, therefore, the duty of the English foreman to induce them, first of all, to stand to their work, and then, to teach them the use of European tools. This has been accomplished by degrees, and the result has been most satisfactory.[16]

Sometimes the British conceded that an Indian tool, technique or method of working was best and it was continued in the same or modified form. Berkley recognized this early and identified the basis of concession: Indian ways were cheaper and quicker.[17] Most of the British engineers and foremen were practical men.[18] They had railways to build—economically and/or expeditiously—and were not too fussy about how that was accomplished. Accomodation was often the best managerial response in the areas of tools and techniques or even, as in the case of earthworking, to an entire labour process. Indians, therefore, shaped and even determined many of the specific work processes. Examples abound: earth continued to be carried in head baskets; Indian divers working in the traditional fashion continued to be used to excavate shallow wells for bridges; stone Wudders changed little their methods of quarrying; clay country hearths heated the rivets for the Jubbulpore extension bridges; bullock-powered dredges excavated wells at Bezwada. The case of the wheelbarrow, however, best symbolizes 'passive' resistance by Indian workers and British accomodation to it. The wheelbarrow never really caught on in India as a device for effectively moving substantial amounts of earth, rock and ballast. Arthur West stated in 1851 that the wheelbarrow 'after the experiment at Sion marsh, was never adopted in future railway works at Bombay'.[19] Another engineer, Thomas Going, recounts a story so evocative as to suggest some embellishment:

---

expenditure was attributed to 'desertion of imported labour, to combination amongst contractors, and to the destruction of bricks during the rainy season of 1884'.

[16] Brunton, *MPICE*, 22 (1862–3), pp. 466–7. The Indian carpenters use of their toes and feet fascinated the British. See also Berkley, *MPICE*,19 (1859–60), p. 609.

[17] Berkley, *MPICE*,19 (1859–60), p. 608.

[18] Glover's diatribe against supervising engineers notwithstanding. See chapter three.

[19] IOL&R, Mss. Eur. D. 1184, West Memoirs, part II, entry for 3 March 1851. Nor, for that matter, were big spades though for a somewhat different reason. The lightly shod (if at all), often frail Indian coolie was not suited to wield the kind of

> It has been often attempted to introduce the wheelbarrow mode of work, but with little success. The basket of antiquity—probably antediluvian—still holds its own. I have heard of an instance of an enthusiast in wheelbarrows who, having exhausted his morning energy in the fond endeavour to restrain a gang of coolies from using the objectionable basket, had the mortification, on making his evening tour of inspection, to find them carrying the wheelbarrows on their heads, in the belief that it was only a convenient modification of the principle.[20]

Macgeorge wrote in 1894 that mechanical appliances were little used in earthmoving in Indian railway construction and that the material was carried in little baskets on the heads of men, women and children.[21] Why then was the wheelbarrow not used more extensively? Because Indians did not want to use it and resisted. The British acknowledged the resistance (Going's anecdote may be humorous, but it also concedes defeat) and acquiesced in the continued use of the basket because they, in turn, discovered they could cut, embank and ballast expeditiously and cheaply using the Indian method. Why the Indians resisted in this area (while Indian carpenters adapted in others) leads us into areas beyond the concerns of this chapter and into the sources of worker agency. One reason for resistance, however, was precisely because many men, women and children could be employed under the existing system. Davidson hit the nail on the head: 'It allows, too, the whole strength of a family to be employed, from the grandsire down to the girl and boy of ten or twelve'.[22] Many families counted on the participation of most family members to make a living wage for the entire family. There was nothing irrational about the resistance of Indians to the wheelbarrow. There was something unjust about a system that required the labour of an entire family to earn a subsistence wage. Possibly, too, male heads of households did not want

---

spades use by the brawny, heavy-booted British navvy. There were exceptions. Wheelbarrows were used during the building of the Ganges Canal. See Lt. Colonel J. G. Medley, compiler, *The Roorkee Treatise on Civil Engineering in India*, 3rd ed. (Roorkee: Thomason College Press, 1877), vol. I, p. 243.

[20] IOL&R, Mss. Eur. C. 378, typescript copy of 'A Pioneer of the Madras Railway, 1867–1875', based on the diaries of Thomas Hardinge Going (1827–75), p. 10.

[21] G. W. Macgeorge, *Ways and Works in India* (Westminster: Archibald Constable and Company, 1894), p. 327. However, at one large embankment on the Thal Ghat 27 tramways were used to help move earth. A labour saving device like a tramway was more likely to be used at a long term project but in the case of the Thal Ghat incline labour shortages may have contributed to the decision to use more capital-intensive methods.

[22] Edward Davidson, *The Railways of India* (London: E and F.N. Spon, 1868), p. 162.

to use the wheelbarrow because they would have had to work harder since wheelbarrows would have been too heavy for the women and children. Much more needs to be known about gender-based divisions of work among construction workers, especially since much unskilled railway construction work utilized family labour.

Collective action took a variety of forms and was sparked by a variety of causes. The predominant cause was wages: their method of calculation, their amount and their full and regular payment. Irregular payments, or payments less than what the workers believed had been promised, were most likely to move the workers to strong action. Resistance to the introduction of piece-work was muted and ambivalent. Braverman states: 'Piece rates in various forms are common to the present day, and represent the conversion of time wages into a form which attempts, with very uneven success, to enlist the worker as a willing accomplice in his or her own exploitation'.[23] A Madras engineer called it a method of securing 'the self-interest of labour'.[24] An example of such co-option occurred in a BB&CIR brickyard in the 1850s. The same workers who had initially resisted the introduction of the heavier and more demanding brass mould, struck work when, due to shortage of the new moulds, they had to work with wooden moulds, and found that 'they could not earn within two annas per day of their more lucky brethren' who were making over 400 hundred more bricks per day.[25] From one perspective, the brick-makers can be viewed as accomplices in their own exploitation. The terrain of compromise basically worked to managerial advantage since the brass moulds offered the lure of higher pay.

A major disturbance on the Bhore Ghat works in January 1859 well illustrates the explosive potential present in irregular and short payments.[26] Working and living conditions on the Ghat were harsh at all times, so when the sub-contractors, some of whom were months in arrears with their wage payments, began to pay out on 17 January at half the promised rate, the workers became angry. By 20 January 'the men had prepared themselves by arming with sticks and stones and eventually they attacked the Europeans'.[27] Later, armed Europeans went to the coolie

[23] Braverman, pp. 62–3.

[24] IOL&R, L/PWD/3/219, Madras RR Letters, enclosures to no. 7, dated 16 June 1876.

[25] IOL&R, L/PWD/2/121, 'Memoranda for those engaged in construction'.

[26] Much of what follows regarding this disturbance is drawn from my 'Working Class Protest in 19th Century India. Example of Railway Workers', *EPW*, 20:4 (26 January 1985), p. PE 35.

[27] MSA, PWD (Railways). 1859, vol. 25, compilation 215: 'Bhor Ghat. Disturbances among the workmen employed on the'.

huts to arrest the ringleaders. One European who had wandered away from the armed party was later found dead—shot through the head, but by whom and how was never uncovered. The riot, sparked by the wage issue, undoubtedly represented the culmination of a festering situation on the Ghat. The coolies lived in an isolated and difficult environment. Petty violence and intimidation towards Indians by lower-level Europeans was endemic. One senior British official observed that it was a wonder, given the conditions on the Ghat and the undoubtedly shameful treatment of the labourers, that an outbreak had not occurred sooner. Moreover, he said: 'It is evident that the labourers have been most grossly abused in the matter of their wages'.[28] The riot was not an aimless, mindless affray. The central grievance was clearly wages. The rioters had ringleaders; they had the Europeans as a focus for their animosity; and they had made some preparations. Many of the labourers were tribals, among whom were the Mhangs and Minas, whose muccadums probably were instrumental in focusing on and articulating the grievances of their gangs, both before and after the riot. This suggests that pre-existing patterns of leadership and identity were important in assisting these workers to act together. The outcome was a partial victory for the workers. They got the close attention of the government. An investigation was launched, pressure was applied to ensure prompt wage payments, and the police and magisterial presence on the works was increased. The British response, therefore, was to try to reduce some of the causes of worker discontent and, at the same time, to enhance official and managerial control. Conditions did improve, but the harsh life on the Bhore Ghat and on the Thal Ghat ensured that both continued to be scenes of unrest until the completion of the railway inclines.

Another case of violence against Europeans occurred in March 1878 in Madras at the Vellore Girder Bridge works.[29] Thirty coolies of the Pariah outcaste rushed upon an engineer assistant named J.D. Crabbes 'and assaulted him most brutally with heavy wooden mortars killing him on the spot'.[30] The Pariahs were reported to have been the most rebellious in the camp and they 'had always resisted the authority of Crabbes'.[31] The cause of the attack was not specified. Direct attacks on Europeans were rare in railway construction, much rarer than Europeans hitting Indians, though this restraint on the part of the Indians can be partially explained by the limited and infrequent contact between any *particular* gang of

[28] Ibid.

[29] Kerr, *EPW*, see fn. 26.

[30] *Madras Times*, 19 March 1878, p. 2.

[31] Ibid.

Indian workers and its European supervisor. Perhaps Crabbes was inclined to abuse his workers physically. There is merit in Chakrabarty's idea that the form of worker protest was related to the way in which they experienced authority.[32] Burawoy goes further and writes about a production regime he labels 'colonial despotism: despotic because force prevails over consent; colonial because one racial group dominates through political, legal and economic rights denied to the other'.[33] Brutal managers tended to spark worker vengeance, but this occurred on railway construction only in a few cases where the European presence was unusually palpable, prolonged and abusive. Nonetheless, managerial brutality does appear to have been the deciding factor in pushing some workers beyond peaceful protest and into violence. Wage issues brought the workers to collective action; brutality pushed them into making that action violent.

The latter point is illustrated by the situation in Sind close to the time of trouble on the Bhore Ghat.[34] Edwin Bray suddenly abandoned his contract and left for England, leaving some 12,000 workers unpaid. The workers went on strike and refused to go back to work until they received what was owed to them. Some said they would not work on the line again, and would return to their villages once they were paid off. Others complained about having had to pledge their clothes to banias in order to get food. The investigating official concluded that payment of wages in full was necessary if the work was to get going again and that if the system of keeping the men in arrears continued 'it will eventually lead to some serious breach of the peace. The Jussulmeer men in particular do not appear to be men that will stand being trifled with'.[35] The reference to the Jaisalmir men suggests that the subjective state of these men, based perhaps on their perception of a martial past, shaped their response to this particular situation. The Jaisalmiris were perceived by the British official to be more volatile than other workers whose objective condition was the same. Interestingly, another group of workers on the Bray contract,

[32] Dipesh Chakrabarty, 'On Deifying and Defying Authority: Managers and Workers in the Jute Mills of Bengal 1890–1940', *Past and Present*, no. 100 (August 1983), pp. 124–46. In the jute mills, workers and European managers and foremen were in much more regular and sustained contact.

[33] Michael Burawoy, *The Politics of Production* (London: Verso, 1985), p. 226.

[34] Kerr, *EPW*, p. PE 36.

[35] MSA, PWD (Railways), 1859, vol. 60, compilation 327: 'Sind Railway. Relinquishment by the contractors of the railway works and arrangement for their construction departmentally.'

200 Kutch masons interviewed by the same official, stated they had struck not so much because of the pay arrears, which they believed would eventually be paid, 'but because the contractor's agent had placed a Jussulmeer Muccadum over them under whom they refused to work'.[36] One needs to be sensitive to the fact that economistic explanations do not satisfactorily explain all instances (or all facets of a given instance) of worker resistance, either with respect to cause or to object. Ranajit Guha reminds us of this when he writes that Santhal attacks against railway works in 1855 were sparked by a concern for the honour of their women and their own dignity, 'against the raping, bullying railway sahabs. In other words, in one of those unpredictable leaps of consciousness prestige suddenly assumed for them an importance exceeding that of money and politics transcended economics'.[37] The Railway Company finished the work in Sind departmentally. Labour peace was restored, and an adequate supply of labour ensured by guaranteeing daily wage payments.[38] The wage rates under the new system were 25 per cent lower than Bray's rates. Clearly, for the workers, assured payment was more important than the higher rate—perhaps because they were close to the margin of subsistence and could not risk uncertainty.

Short or late payment of wages was something that often happened where work was being done by contract. This practice was most frequent in the 1850s and 1860s when a number of men who were either disreputable and/or had neither the knowledge nor the financing to undertake railway construction in India, took contracts or sub-contracts. Examples of worker action caused by arrears in wage payments or short payments rarely appear in the records in the later decades, and that absence probably corresponds to what actually happened. Like everything else in Indian railway construction, contracting became more routinized and effective as the decades passed. Engineers, taught by experience and precept, became more vigilant in making sure that workers were paid on time.[39] Contractors and petty-contractors became more disciplined. One of the few late examples I encountered of wage arrears—ironically where the

[36] Ibid.

[37] Ranajit Guha, *Elementary Aspects of Peasant Insurgency in Colonial India* (Delhi: OUP, 1983), p. 143.

[38] Brunton, *MPICE*, 22 (1862–3), p. 455.

[39] E. Monson George, *Railways in India* (London: Effingham Wilson, 1894), p. 49 states: 'The engineer if he treats his men fairly, and sees that they are paid regularly will have no difficulty in getting what labour he requires to execute his work cheaply, expeditiously and well.'

work was being done departmentally—was a mild collective protest in the quiet voice of a petition. For obvious reasons petitioning was not much used by construction workers. However, in 1890, 17 coolies at work on the SIR in North Arcot district sent a petition to the Madras Government which stated that the engineer was not paying them fully and promptly.[40] Their words, even if written for them, echo in my ears: 'The coolies who are at work in such hot days, suffer from starvation. Would their starvation and wages in arrears produce any profit to the Government? It is divinely charitable, if instead of forcing or at least killing the coolies . . .' they get paid regularly. 'The poor suffer and starve and the fellow creatures enjoy and are surfeit.' The petition prompted an investigation and the engineer was instructed to make sure the coolies were paid more regularly.

One unusual example of collective action deserves mention. Well-paid Punjabis engaged in sinking wells for the Godavery bridge at Rajahmundry in 1897–8 combined to make the job last as long as possible. This sophisticated attempt to control the work process—an attempt possible in construction work only when a small, homogeneous group of skilled workers was engaged in the same task—was recognized by the engineer and defeated by the decree that if a day's sinking in rock was less than six inches, a general reduction of pay would take place.[41]

The main form of on-the-job collective action taken by the workers was the strike.[42] Strikes were not uncommon in any of the decades covered in this study, and were more frequent than the limited examples that survive in the records suggest. Berkley wrote in 1859 that strikes 'have, occasionally, taken place', while another engineer at the century's end wrote: 'There are no trade unions in India; still, strikes are by no means uncommon, as every engineer knows.'[43] Many strikes undoubtedly were of short duration—a temporary downing of tools—and did not impede work sufficiently to enter the records. Others entered the records because of the extraordinary action taken to break the strike. Such, for example, was probably the case with a strike recorded on the Tapti viaduct of the

[40] TNSA, PWD (Railway), 19 May 1890, no. 690 MR (B).

[41] India, Director of Railway Construction, Technical Paper no. 86: F. T. G. Walton, 'The Construction of the Godavari Bridge at Rajahmundry on the East Coast Railway' (1900), p. 4.

[42] How many people simply found railway work too hard, too distasteful or too poorly paid and left cannot be determined though they also were registering their form of protest. And then there were those absconding coolies mentioned elsewhere in this book who 'manipulated' the system of advances to their advantage.

[43] Berkley, *MPICE*, 19 (1859–60), p. 606; J. W. Parry, 'Some Features of Indian Railways', *Engineering Magazine* (July 1898), p. 571.

BB&CIR in February 1860.[44] It was claimed that the coolies struck, despite the notice that rates of pay were to increase, and the engineers quickly struck back. One engineer went to Baroda and collected several hundred men to do the work of the strikers. Others followed and 600 Baroda coolies were soon at the site and the strike was broken.

Significantly, most of the recorded strikes in the later decades took place among the skilled workers whose more limited numbers facilitated combination, and whose more critical role at certain stages in a work process provided greater leverage. Also, their greater subordination to capital may have heightened their proletarian consciousness. Earthwork coolies had to strike *en masse* to be effective; whereas, relatively few skilled workers, if they chose the right moment, could bring work to a halt at a crucial juncture (e.g. assembled but loosely bolted bridge girders were vulnerable until riveted). Three hundred riveters struck in August 1880 at the PNSR's bridge across the Sohan, but returned to work three days later.[45] Riveters also struck in April 1903 at the Bengal–Nagpur's coal-line bridge across the Damodar.[46] Divers struck for higher wages during the construction of the IVSR's Jhelum bridge in the early 1870s and for the same reason on the PNSR's Harro bridge in the early 1880s.[47] A nearly successful strike for higher wages among skilled workmen on the Sind–Peshin line in 1889, was put to an end by the arrival of a contingent of Punjabi workmen obtained by a recruiting engineer despatched for the purpose. This enabled the engineer incharge to stick to the 'rates of wages I should otherwise have had very largely to increase'[48]. Circulating labour again ensured labour discipline and low wages.

The heterogeneous and mobile character of the railway construction workforce as well as the nature of the work made it impossible for the workers to combine widely for purposes of collective action. Organized resistance was always restricted to the locality and to the individual worksite, even to particular gangs of workers. Inchoate forms of passive resistance, e.g. to the wheelbarrow, could be more widespread since they tapped some deeper shared concern. The precarious situation of many

[44] MSA, PWD (Railways), 1860, vol. 54, compilation 118: 'Progress Reports furnished by the BB&CI Railway Company'.

[45] IOL&R, P/1704, PWD, Railway Construction Progs., March 1881, no. 98.

[46] Taylor, *MPICE*, 155 (1905), p. 323.

[47] IOL&R, Mss. Eur. D. 904, Rayne Papers, Diary entry, 23 June 1872; John Willcocks, 'The Harro Bridge, Punjab Northern Railway', *Professional Papers on Indian Engineering*, 3rd series, vol. I (Roorkee: Thomason Civil Engineering Press, 1883), p. 30.

[48] IOL&R, PWD, General Progs., 18 January 1890, nos. 38–42.

workers also made it difficult for them to sustain strike action for an extended period of time. Those who could strike longest were those least in need of railway work to survive. Nonetheless, despite the many obstacles they faced, Indian workers did resist certain aspects of railway construction work and they did so in an impressive and diverse set of ways.[49] Or, more positively, they sometimes seized an opportunity where they had some leverage to try to improve their material lot. Usually, however, the attempts at resistance or improvement failed. Reservoirs of unemployed labour ensured that in all but the short run, collective action could not be sustained. The increasingly extensive labour markets came to serve capital well. Construction workers in nineteenth-century India, like those in the 1990s, were a particularly oppressed and powerless body. But they did try. Not because of any criticism or understanding of the great transformation in which willy-nilly they were caught up, but because they wanted to do something about the specific experiences of exploitation and oppression they encountered at the work-sites—sites already too full of danger and misery to have the added component of human exploiting human.

Some engineers admired their workmen and praised their tenacity, fortitude, ingenuity, hard work and willingness to learn. But, at the bottom, the relationship between the British and their Indian workmen was an instrumental one. And, in so far as the resistive attitudes and actions of Indian workers were concerned, they were, for the British, the human counterparts of the physical features of India: obstacles to be overcome, surmounted, modified or bypassed in the effort to build the railways in India. Some of the ways in which the British dealt with specific instances of resistance were mentioned in the preceding discussion. The task that remains is to sort out the levels and varieties of British response and to ask whether the British response to challenges to their authority changed over the decades covered in this study.

In so far as the refusal of Indians to work at all on railway construction was concerned, this is an issue that lies more in the sphere of mobilization of labour, examined in Chapters three and four. Moreover, as also demonstrated in those earlier chapters, the supply of labour ceased to be

[49] One suspects that construction coolies, too, occasionally expressed their anger through attacks on the operating railway lines through robbery or sabotage of the track. For such activities among the permanent way maintenance staff of the operating lines read Dipesh Chakraborty, 'Early Railwaymen in India: "Dacoity" and "Train-Wrecking" (*c.* 1860–1900)', in *Essay in Honour of Professor S.C. Sarkar* (Delhi, 1976), pp. 523–50.

a major issue during the 1860s. No doubt, Indians continued to refuse to enter the construction labour market but they were of no significance to an engineer who had the labour force he needed. What is of interest is the extent to which British engineers used, or countenanced in others, forms of impressed labour in the 1850s and 60s—with government turning a blind eye—to overcome labour shortages. One such situation was described in Chapter four during the EIR construction in sub-division Barh but, on the whole, I came across few unambiguously specific references to such impressment. Henfrey's description of Punjabi villagers being driven from their home in the morning may be a hyperbole. It is not an issue easily clarified, especially since extra-economic relationships between the workers and those who commanded their labour existed extensively in the railway construction workforce.

Worker resistance of the more inchoate and 'passive' variety, such as opposition to a particular tool or a particular way of working, was met by a British response which became either increasingly clever and successfully manipulative, or increasingly accomodative as the half century passed by. Managing engineers chose one or the other strategy—or both, since they were not mutually exclusive—for reasons which were situationally specific either with respect to the location, conditions and kinds of work, as well as to the background and the abilities of the engineer involved. Regardless of the chosen strategy, the most important point is that the British and their Indian subordinates in the supervisory chain learned to deal with certain kinds of worker resistance and to adapt to or overcome that resistance in ways which were not directly confrontational. Getting the work done became much more important than getting the work done in a particular way. This doctrine was applied clearly by the engineers who built the hill section of the Assam–Bengal Railway (1896–1904). They utilized Indian tools and appliances whenever possible 'and where they did not affect the progress of construction, native customs of labour or even prejudices were not interfered with'.[50]

Strikes were a different matter. The attitude and actions of British engineers and supervisors towards striking workers changed little between 1850 and 1900. Strikes had to be crushed, and striking workers put back to work or replaced. The workers' demands, which most often involved

[50] Nolan, *MPICE*, 178 (1909), p. 322. Cf. this attitude with that expressed by the young Fowler in the 1850s described in chapter two. Nolan, like Fowler, labelled certain Indian attitudes and practices as 'prejudices' but Nolan and his fellow engineers had moved an enormous distance in their willingness to accomodate themselves to those 'prejudices'.

the size or the security of their pay, were met only when no other choice was available. Usually, the engineers had other choices. Some work processes could be suspended for a period of time without jeopardizing the progress of the overall project. Workers rarely had the resources to sustain a long strike, and not all of them could simply exit. More commonly, the engineers arranged to hire replacement workers from the increasingly available supply—a move which rapidly brought most strikes to an end. Only the specially skilled workers acting together at a crucial stage in a construction project, were temporarily immune to the threat of replacement.

The partial exception to the scenario sketched above was the very large strike, possibly accompanied by violence, which drew the government's attention. India's colonial administrators—especially in the aftermath of the events of 1857–8—did not want massive unrest at construction sites. They also wanted the railways built. Unrest on the Ghat inclines and in Sind resulted in government investigations which partially vindicated the workers. Particularly after the violence on the Bhor Ghat, the colonial government became more inclined to ensure peace at work-sites through the intensified presence of magistrates and the police. The Government of India, always extensively involved in the construction process and with its own particular interest in seeing projects finished, ultimately ensured labour peace.

The Bhore Ghat violence of 1859 led to the introduction of a bill in the Legislative Council of India, 'to empower Magistrates to decide disputes between contractors and workmen engaged in railway and other public works'.[51] This bill became the Employers and Workmen (Disputes) Act (X) of 1860, in which magistrates were given summary powers to settle wage disputes. Interestingly, an Act initially proposed because of a clear case of maltreatment of workers, also came, in the course of its formulation, to include provisions for fining or imprisoning workers who, having engaged to work for a particular period or carry out a specific work, failed to fulfill their commitment. No appeal was permitted against a decision passed under the Act.[52] However uncertain the enforcement may have been, the colonial state began to weigh in on the side of capital. The railways of the Raj, it must not be forgotten, were built with and through the close involvement of the colonial Government of India which was not

[51] *Proceedings of the Legislative Council of India, for January to December 1859*, vol. V (Calcutta, 1859), p. 217.

[52] The Act is reprinted in Henry Edward Trevor, *The Law Relating to Railways in British India* (London, 1891), pp. 347–8.

a neutral, uninterested party standing above the construction process. A legal framework suitable for the advance of capitalism, one which reflected the subordination of labour to capital, began to take shape.[53] This subordination came to be directly expressed in a series of major railway acts (1854, 1879 and 1890) whose labour sections were directed primarily at the employees of the operating railways, but whose ambit included permanent employees engaged in construction or reconstruction. Engineers and contractors had a formidable ally when it came to controlling their workers, even though that ally would sometimes, in specific situations, side partially with the workers.

[53] For further explorations of this process see Jan Breman, *Labour Migration and Rural Transformation in Colonial Asia* (Amsterdam: Free University Press, 1990), pp. 60–9 and Burawoy, *The Politics of Production*, p. 214 ff.

CHAPTER 7

Conclusion

No trains ran in India in 1850. Twenty-five years later, trains ran over a railway network encompassing 6541 route miles. The network grew to an impressive 23,627 miles by March 1900, at which time an extensive system of trunk lines and many branch lines criss-crossed the subcontinent.[1] A traveller, at the turn of the century, could depart from Bombay's magnificent Victoria Terminus, itself a luxuriant symbol of India's railway age, and be quickly transported to distant cities like Karachi, Peshawar, Calcutta, Madras and Calicut. In reaching these places, the train took our traveller into the kaleidescopic interior of the subcontinent: an ever-changing panorama of people, places and landscapes. The railway went through cities as diverse as Hindu Benares, Mughal Lahore and colonial Bangalore; it passed many of India's half-million villages whose inhabitants and habitations were revealed to the passenger in brief, tantalizing glimpses, from carriage windows; it crossed many landscapes: dry, nearly empty wastes; intensively cultivated wet-rice paddy fields; dense jungles; steep inclines; flat plains; wide rivers. And where the passenger trains went, so too did the freight trains, knitting India's economy into an unprecedented, integrated whole.[2] The railway had possessed India, as it still does.

Our traveller in 1900, probably, took all of this for granted. The train steamed onwards night and day, but the traveller, whether ensconced in the considerable comfort of a first class carriage or in the hard discomfort of a third class carriage, probably gave no thought to the workers whose efforts in the deep cuttings, the high embankments, the tunnels and the bridges, had made steam locomotion possible. As one author put it, in 1894:

> There are probably but few travellers now daily passing up and down the magnificent Thul and Bhore ghat inclines, quietly seated in comfortable railway

[1] 25,101 running miles.

[2] The role of the railway in creating an integrated, India-wide market in food grains has been demonstrated by John Hurd II, 'Railways and the Expansion of Markets in India, 1881–1921', *Explorations in Economic History*, 12:3 (July 1975), pp. 263–88.

carriages, who can at all adequately realise the extraordinary nature of the obstacles which have been so successfully overcome, and the great skill and daring of all those engaged—especially during the first years—in shaping and carving out of the rocky mountain sides those wide luxurious roads on which they now so easily and securely travel.[3]

The railway was there—it was an accepted and integral part of the life and landscape of India. Its construction was the great collective accomplishment of human labour, as detailed in the preceding chapters. Unlike our traveller, we now know something about the builders of the railways; where they came from, how they worked; and, in the particular focus of this book, how the British managed the construction process.

Capital raised largely in Britain, and, later, direct investment by the Government of India funded the construction. Private companies built and owned the railways until 1869; private companies and government built, owned and/or managed railways during the remainder of the nineteenth century. However, because of the guarantee, the Government of India and its subsidiary jurisdictions maintained a close supervisory control over private companies. Government, therefore, was an integral and unifying part of the entire railway construction process in India in the period between 1850 and 1900. During this period, there was, on an average, 1405 miles under construction every year although, as is demonstrated in Table 1, the amount varied considerably from year to year.

The general management of the construction of the railways of the Raj followed two patterns. First, there was the large-contract system. The majority of the large contractors were British, but among them were also men like the successful Jamsetji Dorabji and the unfortunate Muhammad Sultan. The second pattern was the departmental system, in which the engineers of private railway companies or of the State railway lines acted as prime contractors. Both patterns persisted throughout the period, 1850 to 1900, although the large-contract system was more prevalent in the 1850s and the 1860s. Below the prime contractors or the engineers were the intermediaries, largely Indians, who stood between European levels of middle and upper-level management and Indian workers who actually performed the manual labour. These intermediaries can be labelled petty contractors and headmen. The growth of a class fraction of petty contractors made an important contribution to the ability of the British to build railways economically and/or expeditiously. The more numerous presence of competing petty contractors, more knowledgeable in railway

[3] G.W. Macgeorge, *Ways and Works in India. Being an account of the Public Works in that Country from the Earliest Times up to the Present Day* (Westminster: Archibald Constable and Company, 1894), p.358.

construction and with a better understanding of what it meant to take and fulfill a contract, made it easier for the British to obtain labour and be less concerned about the way in which that labour worked. The advance of capitalism disciplined both labour and the emerging petty bourgeoisie. Headmen and headwomen (gangers, muccadums, maistries) were lower still, in the chain of intermediaries; they sometimes functioned as sub-petty contractors.[4] The intermediaries were, for the most part, the ones who commanded labour; they mobilized and controlled the construction workers—in ways that included extra-economic links and sanctions. The petty contractors had limited capital resources of their own, and they depended on advances and interim payments from large contractors or engineers, to obtain and retain workpeople. These advances represented the advance of working capital from the controllers of large sums of capital, to the most petty of capitalists. The advance linked British capital to those who actually built the railways; the advance, through one or more layers of intermediaries, purchased the labour power of Indian workers. The system of advances was important to the mobilization and retention of labour.

The British were the overall managers and the technicians of the construction process. They provided the know-how and co-ordination that enabled the construction to proceed. Construction involved three major tasks: the formation of the line; ballasting and laying of the permanent way; and the building of workshops, stations and so on. Many specific work processes had to be co-ordinated to complete these tasks; most of those processes involved the use of large gangs of labourers. Some of the processes were new to the Indians who undertook them; other processes, especially earthworking, represented a pattern and a rhythm of work similar to that practised in India for centuries, or longer. The British, however, found these enduring forms of work cheap and effective, while Indians adhered to them because they provided for the continuation of family labour in a context of low wages, and, quite possibly, maintained gender-based divisions of labour within family units, which favoured adult males.

As the decades passed, engineers came to understand better the technical and social requirements of railway construction in India. The railway construction process became routinized. Projects were completed more quickly. The management, increasingly, acquired a better control of the overall process of construction. This did not mean that British engineers

[4] Although there may have been exceptions, I am persuaded that headmen were present in most cases where the 'day-labour' system was in use.

necessarily instituted tighter control over the point of production. Sometimes they did, but what was equally important was the British recognition that limited interference and flexibility in the face of resistance or presence of effective Indian work practices, was often the best way to get a project finished. The presence of better and more numerous petty contractors facilitated the growth of that understanding. The technical content (tools, work skills, etc.) of many individual labour processes, especially the huge labour-devouring category of earthworking, remained substantially unchanged, although change did occur in other areas, such as brick-making and bridge-building.

Thus, the first answer to the question, how did the British get the railways of India built, is: through the use of Indian labour and Indian intermediaries who supplied the labourers and controlled them at the point of production. The second answer is that as the years passed, the British themselves got better at managing Indian intermediaries and their workpeople. Increasingly, the British became better upper-level managers of the construction process, precisely because they learned, adapted and transmitted their acquired knowledge to future generations of engineers. In the context of nineteenth-century India, the acquisition and routinization of knowlege was a powerful force in the increasingly smooth advance of railway construction. The transition from the mistake-ridden start-up years of the 1850s and early 60s, to the settled procedures of the 90s, was evolutionary, but the overall change was considerable. To capture the full force of this transition, one needs only to compare the statement of G.W. Macgeorge, quoted in Chapter one, where he describes how little was known and available in the first decade of railway construction, with the following statement written in 1902:

> We have had considerable experience of construction on the Bengal–Nagpur Railway; and to facilitate and assist in building our extensions, we have had printed what we call *The Construction Manual.* This book informs the staff of the method of organization and what they may and what they may not do. In conjunction with plans and drawings supplied them, and the fixing of rates that may be paid for work, an officer knows what he may do and is in a position to start work with clear directions.[5]

Construction employment averaged some 180,601 to 221,253 persons per year from 1859 to 1900, based on an estimated employment of 126 to 155 persons per mile, and a construction period averaging 2.5 years

[5] (Sir J.R. Wynne), *Notes on the Construction of Railways in India for Students* (Calcutta: Bengal Secretariat Press, 1902), p. 6.

per mile. Table 2 summarizes these findings. Reconstruction employment, if it could be estimated, would add significantly to these figures, as would the immense effort required to get construction materials to work-sites. And, since turnover within the railway construction workforce was high, many more individuals were employed in the course of a year, than the figures in Table 2 suggest. Roughly 80 per cent of the workers were unskilled. Women and children were employed extensively. All able-bodied family members, in many cases, had to work to ensure their mutual survival. Unskilled labour was recruited primarily from low-status castes and tribes, most noteworthy among whom were the Indian 'navvies': the traditional earthworking groups, such as the Wudders, the Odhs and the Nunias. Skilled workers came from various backgrounds, including the traditional artisan castes. Railway construction also created its own pool of skilled labour, through on-site instruction provided by skilled British workmen and engineers, in the early decades: from which emerged a pool of self-reproducing, skilled Indian workers who subsequently moved across India from one work-site to another. These skilled workmen, as suggested in earlier chapters, became more thoroughly subordinated to the dictates of a labour process under capitalism.

Obviously, the development of a pool of skilled labourers was essential to the expeditious construction of the railways. Equally important, however, was the way in which the British used the traditional earthworking castes and tribes. This study has provided ample evidence of the extent to which these people became an important component of the circulating labour that crowded upon the work-sites. It did not require a great deal of insight on the part of British engineers and contractors to recognize the value of these people as particularly effective earthworkers in their own right, and to see them as pace-setters who established output norms at particular wage levels, which could then be applied to other earthworkers. The fact that they expected to be paid by the piece or the task, further enhanced their value. Once the British 'discovered' the Wudders, the Nuniyas and others, they went to some lengths to employ them on railway projects. All earthworkers at a particular site did not have to belong to a traditional earthworking group—but their presence set an example and a pace of work. I suggest, therefore, that another part of the answer to the question of how the British got the railways of India built expeditiously and/or economically, lies in the existence of a body of circulating labour, prior to the railway age, who could be comfortably incorporated into the railway construction process under conditions of the formal subsumption of labour, and who knew well how to move earth and rock at output levels that the British considered good.[6] The contractor, Glover, put it best when he said that he depended primarily on the Wudders, since local labour had to be trained afresh in every district.[7] Thus, for many of the important tasks of railway construction, the British could focus on the conception and co-ordination, and be engaged in limited, and often indirect, supervision at the point of production. Their supervision may have been overly lax in the early going, but, there too, the British soon learnt the importance of making sure that finished work met specifications. Assessing the outcome was much more important in many construction tasks than changing the way in which it was done.

Regional and inter-regional labour markets for construction labour emerged as the result of the demand of the railways (and of other construction projects) for labour. Labour responded to this demand and became increasingly mobile, and moved to construction work on a semi-permanent or seasonal basis. Operating railways facilitated the movement of this labour. The railways, as a consequence, became increasingly freed from their reliance on construction workers recruited from the less reliable and less 'free' local pools of agricultural labourers. Circulating labour was easier to control, and the complex mixtures of peoples, drawn from different regions and communities, present at major work-sites, made it difficult for workers to combine in order to resist. In the longer run, circulating labour also kept wages depressed because of the unemployed or under-employed status of the groups from which such labour was drawn. The British knew that well, as has been witnessed by their recruitment of replacement workers when strike action was threatened. There were no longer-term shortages of unskilled construction labour. Most construction activity remained labour-intensive throughout the period studied, and beyond.

Work at the construction sites presented a mixture of old and new forms of work. Working conditions, however, were uniformly difficult for most workers. Accidental death and injury was common, but it was disease—cholera, malaria, pneumonia, typhoid fever—that was responsible for the high morbidity and mortality rates among construction

[6] Here and elsewhere in this conclusion (and hence also in the substantive chapters), I provide support for positions, argued more generally, in the insightful pamphlet by Jan Breman, *Labour Migration and Rural Transformation in Colonial Asia* (Amsterdam: Free University Press, 1990).

[7] Although not a Glover contract, something of what that training involved was experienced by the tribals hired at the Nagarunjöo Pass in the late 1860s. See chapter four.

gangs. Workers, too, suffered from oppression and exploitation. Wages were low, life hard, even brutal, and prospects for a better future extremely limited.[8] Workers responded to these conditions in a variety of forms of collective resistance, ranging from the most indirect and passive ones, to strike action and violent outbursts. Though heterogeneous and mobile construction workers could not combine widely, they could, and did, act when conditions became intolerable or when they saw an opportunity to improve their conditions at specific work-sites. The examples of collective action presented in Chapter six did not represent any conscious criticism on the part of workers, of the process of proletarianization into which railway construction work had partially or fully ensnared them. The workers resisted threats to their livelihood, represented by short, irregular or non-payment of wages, or they struck back at those in authority when working conditions became intolerable.

Broadly speaking, the labour process approach adopted in this book, pays particular attention to the role of the British as the managers of the railway construction process. Management is a crucial component in the relations in production, perhaps because initiative usually lies with the management. The railway construction process was a labour process writ large, a complex inter-continental and sub-continental assembly of mental and physical work activities. Within a shared, overarching structure of conception, capital investment, co-ordination and supervision, at whose centre was the Government of India, there existed an assembly of work processes each possessing its own configurations of authority, ways of organizing work, work techniques and tools. Compare, for example, the descriptions and illustrations, in previous chapters, of earthworking, well-sinking and the model brick manufactory advocated by the CE of the BB&CIR. British engineers remained the dominant loci of managerial and technical authority. Indians constituted the overwhelming majority of manual workers.

The simultaneous existence of multiple kinds of work processes and of different expressions of managerial authority in railway construction, supports Burawoy's contention that different expressions of the labour process are compatible with capitalism. Railway construction in nineteenth-century India exhibited, in many different ways, the presence of a labour process under capitalism, rather than the emergence of *the* capitalist labour process following the inexorable path which, as Marx

[8] For most workers, the wages were at best, a little more than those required for subsistence, and this state of affairs became more pronounced in the later decades after the initial wage increases of the late 1850s and the early 1860s.

believed, led from the formal to the real subsumption of labour under capital. The combined and uneven advance of the forces and relations of capitalism in nineteenth-century railway construction in India, exemplified the complexities generally present in the advance of capitalism everywhere. In India, generally, and in Indian railway construction in particular, the advance of capitalism in the nineteenth century depended heavily, as it did elsewhere, on the physical activity of a great number of human beings whose presence provided an abundance of exploitable labour. Moreover, in Indian railway construction, as more generally in industrialization in India or even in Victorian England, productivity could be increased without rapid mechanization.[9]

The British had a constant goal, to get the railways built—and that fixedness of purpose was itself important—in the pursuit of which they used whatever means and people worked best in particular situations. The organization, supervision and conduct of the many construction work processes was determined expediently and existentially within an evolving context of learned experience. As Braverman's theoretical statement suggests, the British did have to deal with the historically evolved consciousness of the Indian workers, and with the general and particular social and technical conditions within which the enterprise of railway construction in nineteenth-century India took place.[10] Therein lay the dynamic interaction between British goals and expectations, technical change and Indian realities, with the result that the best engineers were those who understood best the Indian realities, and who adapted their exercise of authority accordingly. The effective exercise of authority was measured against results—expeditious and/or economical construction; it was measured against what was done, and not by how it was done.

The British did impose a certain technology on India: the railway itself. That technology established certain limits with respect to what had to be constructed, while the physical conditions of India, sometimes, dictated what form a particular aspect of construction might take. Iron bridges, for example, proved best suited for many of India's rivers. The British,

[9] Many of Raphael Samuel's explanations for the slow mechanization of mid-Victorian England, could be applied to railway construction in India, e.g., improved tools within what remained basically a hand technology, or the more systematic exploitation of labour through the manipulation of wage-rates, or the better co-ordination, organization, and division of work activities. See R. Samuel, 'Workshop of the World: Steam Power and Hand Technology in mid-Victorian Britain', *History Workshop*, 3 (Spring 1977), pp. 45–51.

[10] Harry Braverman, *Labour and Monopoly Capital* (New York: Monthly Review Press, 1974), p. 57. I have quoted the statement in question in chapter one.

however, learned to be flexible when it came to questions of how something should get done. If forms of the labour process which Marx would have labelled formal subsumption, remained common, it was because they got the job done. There was nothing special or unusual about formal subsumption in nineteenth-century Indian railway construction. It was an expedient response of management to the conditions of specific times and places, as it has continued to be up to the present.

Effective managerial control can be achieved in a variety of ways. If I read later twentieth-century developments correctly, the patterns of control characterized as formal subsumption are found to be more rather than less prevalent, throughout the capitalist world economy. Even if one does not wish to go that far, one certainly cannot consign the vast world of construction labour over the last two centuries, to the dustbin of surviving specialized instances.[11] Indeed, British control, direct or indirect, over the construction process became firmer, precisely because the workers were caught up, to differing degrees, in a process of proletarianization, and thus became increasingly subordinated to capital through their increased exposure to the vagaries of the labour market.[12] Certainly, this book provides a substantial body of evidence which indicates that railway construction in India extended and deepened the presence of capitalist labour markets. The factor that assured worker obedience at work-sites was most often a labour market with an intimidating pool of reserve workers.

The construction labour markets that emerged in India also included elements of extra-economic coercion. Some of those elements, in fact, were central to the operation and integration of the labour market; they represented the persistence of pre-capitalist relations in mobilizing labour and in exercising authority over labour at work-sites. However, we would be wrong, again, to see anything peculiarly Indian in the presence of these survivals.[13]

Skilled workers came under more direct control of engineers. Based on this, one can tentatively advance the argument that the specific forms of the labour process in which they engaged, made them more thoroughly

[11] Both Marx and Braverman assume that the labour process under capitalism must evolve in a particular direction and hence, for them, formal subsumption survives only in specialized instances.

[12] Richard Price, *Labour in British Society. An Interpretative History* (London: Croom Helm, 1986), p. 21, makes this argument in the British context.

[13] Cf. the Indian situation to the discussion in ibid. ch. 5, 'Work and Authority, 1880–1914'.

proletarian. A second tentative proposition supports the first. The limited evidence available suggests that, as the decades passed, skilled workers proved to be the most likely to give combined voice to protest, although the mobility and segmentations of the workforce restricted combination to particular sites. This voice of protest suggests the growth of a shared consciousness among skilled workers. Regardless of these tentative possibilities, the broader point is surely that proletarianization, in its subjective dimension of worker consciousness, is an uneven process which occurs over an extended period of time. The English working class did not evolve overnight; the many fractions of the Indian working class, burdened with the atavisms of a pre-bourgeois past (à la Chakrabarty) and a hostile colonial regime, evolved still more slowly.[14]

What is remarkable is the extent to which railway construction workers, despite all the obstacles they faced, resisted managerial authority. Worker resistance was something that the British had to deal with—be it through accomodation or repression—in their drive to get the railways of India built. And, deal with resistance, they did. The evidence is that not much delay was caused in the building of the railways on account of Indian resistance (except for the much more broadly based instances of resistance in 1857–8). Physical obstacles and worker resistance notwithstanding, India did have the world's fourth largest railway network by the early twentieth century: a network built by Indian labour, under British direction. Management succeeded. The civil engineers of Victorian Britain had one of their greatest accomplishments, although many among them died on the job. Indians, however, did most of the dying, and most of the work. Sheik Sobhan deserves to be remembered as much as does James John Berkley.

[14] I skirt close to a teleology I reject, with my use of the word atavism. Atavism, I suspect, is the rule rather than an exception. When I begin to turn my hopes into teleologies, I remind myself that most academics do not acknowledge that they are workers (however privileged) who sell their labour power for a wage in order to survive.

APPENDIX I

How Many?

This appendix provides a numerical footing for the discussions of labour recruitment, work and working conditions. The attempt to determine how many people were required to build the railways of India in the nineteenth century, runs into difficulties similar to those encountered by the bridge-building engineers of the 1850s and 60s: there is no assured base of knowledge upon which to build—in my case, secure estimates of the size of the construction workforce. The sources are diffuse, scattered and incomplete. Unlike the operating-line employees, about whom tolerably acceptable employment statistics were kept and published annually from 1860, construction workers were counted on an ad-hoc, site-by-site basis.[1] Moreover, the routinization of the construction process took its toll. The extensive, albeit scattered, references to the numbers of construction workers, which dot the records of the 1850s and 60s, are much less frequent in those of the 80s and the 90s: large gangs of construction workers were no longer novel, nor were their collection a major problem, thus they no longer figured prominently in the records.[2] Estimates of the size of the construction workforce must be extrapolated from the available data. But, regardless of these difficulties, the task of estimation must be undertaken. Some estimate—especially when the sources and procedures of estimation are known—is better than no estimate. Like the bridge-

[1] The operating line statistics were collected annually from the railways from which a condensed version was compiled for publication in the annual *Railway Reports*. Those reports in turn provided the material for the useful, annual open-line employment series (categorized by European, Eurasian and Indian), which appears in Morris and Dudley, 'Railway Statistics', pp. 202–5. Some construction workers may have been counted in the earlier returns but there is no way of knowing how many. One envies the historians of railway construction in Britain who have census enumerations of railway construction workers to base their historical reconstruction on. See David Brooke, *The Railway Navvy* (Newton Abbot: David & Charles, 1983), appendix.

[2] Other references were consigned to the category of 'B Proceedings', i.e., matters of routine, not to be printed. B Proceedings were not sent to the India Office, and few B Proceedings survive in the National Archives of India.

building engineer, the historian cannot avoid the task of foundation construction, just because the knowledge base is inadequate. Tables 2–15 have been developed in the course of this discussion.

Turning first to the Europeans employed in India with a direct, hands-on relationship with railway construction, we find the sources to be curiously silent: more, in fact, is said about Indian employment. And, since we know these Europeans to have been few in number, the magnitude of error in the estimates caused by under-counting or over-counting is proportionately greater. On the positive side, error, in this case, matters less than in the case of Indian employment, since less hinges upon the outcome. Collectively, the British were a crucial presence, but precisely how many of them were present, is not a question on which one needs to spend much time.

Table 3 provides the beginnings of an answer to the question of European employment. Apart from showing a growth in the European presence from a miniscule 17 persons in 1850, to 819 in 1859, as the pace of construction quickened, it also suggests a certain ratio of Europeans to miles under construction. In 1855, the employment total was 326 and the miles under construction some 1400, for a ratio of 1 European to 4.3 construction miles. The 819 Europeans and 1061 construction miles, in 1859, provides the closer ratio of 1:1.3. This table may, on the one hand, include some operating line employees, and, on the other hand, it most likely does not include many of the men working for contractors. On balance, Table 3 probably understates the size of the European presence.

Other information provides similar ratios. The 209 Europeans employed on the 1018 miles of the EIR, through Bengal and the North Western Provinces in the period up to 1860, generate a ratio of 1:4.9, although in the Bengal division alone (417 miles), the ratio was 1:3.1.[3] In 1864, twenty-six engineers were authorized for the construction of the 221 miles of the Jubbulpore Extension of the EIR.[4] However, this line was built by a contractor who maintained an 'Engineering Establishment of

[3] IOL&R,P/163/24, India, PWD, RR Branch Progs., December 1860, no. 42, and P/163/23, August 1860, no. 57. These proceedings are distribution lists for a railway medal struck to commemorate the opening of the line to Rajmahal. Eligibility requirements for the medal, plus decisions I made as to whom I should include in my count (Prog. no. 42 lists 260 individuals) probably result in an understatement of the European numbers. It is also likely that the construction conditions in Bengal did require the greater use of Europeans.

[4] IOL&R, P/217/37, North-Western Provinces, PWD, RR Branch Progs., 15 November 1864, nos. 12–14.

about equal calibre and strength', to that of the Railway Company.[5] If we take this statement at face-value, then some 52 Europeans were employed for a person/miles ratio of 1:4.3. In the middle of the 1870s, the State railway system constructed the 498 miles of the Indus Valley line with an authorized European establishment of 136, which translates into a 1:3.7 ratio.[6] Ten years later, the Bengal–Nagpur line had an authorized establishment of 63 for 318 miles of construction, for a 1:5.1 ratio.[7]

An unusually detailed breakdown of the European presence on the various contracts of the GIPR in 1861–2, provides similar results:[8] 23 Europeans were employed on the ten miles of the Thal Ghat; contract no. 12 (90 miles) had 27 Europeans; no. 13 (62 miles), 51 Europeans; no. 14 (109 miles), 21 Europeans; no. 15 (139 miles), 38 Europeans; no. 16 (80 miles), 13 Europeans; and no. 17 (116 miles), 24 Europeans. These figures, respectively, generate men to miles ratios of: 0.4:1 (i.e. roughly one European to each half mile); 1:3.3, 1:1.2, 1:5.2, 1:3.7, 1:6.2 and 1:4.8. The variance is considerable but the majority of the numbers fell in the 1:3–5 range. The special conditions of the Thal Ghat, like those of the Bhore Ghat, demanded a substantial European presence and extraordinarily large numbers of Indian workers. But, it should be stressed, there were always some railway construction sites in the period under study—great bridges, tunnels, other Ghat ascents—that required larger than normal concentrations of European and Indian labour. The presence of these sites warns us of the dangers of a facile application of estimated averages to particular construction conditions.

The data scanned above suggest that usually one European was employed for each 3 to 5 miles of railway construction. The variance is considerable, given the small numbers involved, and some of it undoubtedly reflected differences in construction conditions. Impressionistic evidence also suggests that fewer Europeans were used as the decades passed; the construction process became more routinized and the cost-saving benefits of Indian labour became ever more attractive. In 1859, Juland Danvers stated: 'In addition to the Engineering Staff, it has hitherto been found necessary to send out almost all the skilled labour required for the construction and working of the lines.'[9] In practice, this meant that higher-

[5] Ibid.

[6] IOL&R, P/558, PWD, RR Estabs Progs, February 1874, nos. 3–21.

[7] IOL&R,P/2536, PWD, RR Estabs Progs, July 1885, nos. 44–7. This was an authorized establishment for State construction; the line was eventually built privately, but, I presume, with a similar engineering establishment.

[8] GIPR, *Annual Report of the Company for the Year 1861–62*, pp. 8–12.

[9] *Railway Report, 1859*, p. 30.

level direction (engineers) and lower-level supervision (overseers and inspectors) were in British hands, and that skilled British workmen were imported for tasks like plate-laying. By 1884, however, a major contractor testified before a Parliamentary Select Committee that he had many Indian overseers whom he found to be 'much more industrious, much more steady, much more to be depended upon than English overseers', and that skilled workmen and their foremen were usually Indians as well.[10] The contractor, Thomas Glover, still reserved the superintending posts for Englishmen or Scots, but other evidence shows that Indians were slowly entering the lower ranks of engineering establishments. Babu Ram Gopal Vidyant, for example, served as an Assistant Engineer for 6.5 years on the Oudh and Rohilkhund Railway (O&RR), with special responsibility within the Chief Engineer's office, albeit closely directed by the latter, for the execution of the design drawings for the great Dufferin bridge across the Ganges at Benares. But, after 26.5 years' service with the O&RR (whose initial construction was sanctioned in 1864), Ram Gopal was still 'the lowest graded Engineer on the Staff'.[11] On a wider scale, the PWD employed 63 gazetted Indian engineers in all its grades, out of a total of over 900 in 1888, of whom only 19 appear to have been associated with construction or open-line railway work.[12] The Indianization of the railway engineering staff proceeded slowly; most of the Indianization occurred below the level of professional engineers. By the 1880s, skilled construction workmen were imported from Britain primarily for special situations. One such situation occurred during the excavation of the Khojak tunnel in 1889–91, where treacherous water-bearing strata required large-scale timbering (to hold up the roof of the tunnel before it was lined and bricked), for which some 65 Welsh miners were obtained from Britain.[13]

In addition to the sources used above, my impression, derived from a reading of an extensive body of railway-related sources, is that European

[10] *PP (Commons), 1884*, cmnd. 284, 18 July 1884, *Report From the Select Committee on East India Railway Communication*, paras. 1920, 1938–40, 1955.

[11] IOL&R, P/3457. PWD, General Progs, July 1889, no. 224.

[12] IOL&R, V/12/54/ India, PWD, *History of Services of the Officers of the Engineer, Accounts, and State Railway Revenue Establishments, including the Military Works Department Under the Military Department*, 4th edn, corrected to 31 December 1888 (Calcutta: Superintendent of Government Printing India, 1889). In a quick check of some 900 names and service records mistakes may have been made, so the figures should be treated as good approximations. The conclusion, however, is solid: there were few Indian engineers in the nineteenth century.

[13] P.S.A. Berridge, 'The Story of the Khojak—The Longest Railway Tunnel in India', *Indian State Railway Magazine*, 7:8 (May 1934), p. 413.

employment decreased over the years. Moreover, the more precise sources cited above tend to understate the European presence in the earlier decades.[14] Understatement was particularly likely where much of the construction was done by contractors. I suggest, therefore, for the purpose of developing estimates of the annual, all-India aggregate size of the European component of the railway construction workforce for the period 1859–1900, that one European was employed for each 2.5 miles of construction from 1859 to 1870; one for each 3 miles in 1871–80 and one for each 4 miles after that date. The mechanical application of these ratios to generate an annual employment series, such as is presented in Table 2, most likely overstates the inter-year variation. Most Europeans, unlike Indians, expected to be employed steadily for a number of years. The final column in Table 2, therefore, presents European employment in the form of a three-year moving average, in order to reduce the inter-year variation. The transition to a reduced European presence was gradual. Regardless of all this, the most valuable aspect of the estimates of Europeans is to demonstrate how so many Indians were directed by so few Europeans.

Now to move to the numbers of Indians employed: these, most certainly, weren't small. Hundreds of thousands of Indians were employed annually; aggregated over the entire time period examined in this study, the number exceeded ten million. How does one arrive at such large numbers?[15]

[14] The appeal to an extensive body of railway-related sources, is not very satisfactory. The Proceedings of Government volumes, in particular, produce fragmentary evidence that generate impressions rather than defensible reconstructions. However, an item like James J. Berkley, *An Address Delivered at the Annual Meeting of the Bombay Mechanics' Institution, in the Town Hall, on Saturday, April 11, 1857* (Bombay: The 'Bombay Gazette' Press, 1857), helps to reveal the basis for the kinds of impressions one can acquire. Berkley, pp. 8–12, gives a most useful listing, by enterprise, of Superintendents, Engineers and their Assistants, Surveyors, Overseers, Maistries and Mechanics (19,801) and Labourers (79,756) in the Bombay presidency, 'besides all those employed by Contractors and in private Manufactories and Workshops'. 9001 of the first category were employed by the GIPR. How many more were employed by contractors? And of that enlarged total how many were Europeans? Quite a few more, one suspects, than the 117 listed in table 3 for 1857.

[15] Some of the following material originally appeared in Kerr, 'Constructing railways in India—an estimate of the numbers employed, 1850–80', *IESHR*, 20:3 (July-September 1983), pp. 317–39. I am grateful to the Editor of *IESHR*, for permission to reproduce that material here. In the reconstruction that follows, I draw heavily on that article, although I have revised those earlier estimates, and extended them beyond 1880. I do not reproduce, in full or in part, all of the tables which

The sources provide no extended serial data on the employment of Indians in railway construction. However, the sources do contain some bodies of information, often quite detailed, about the numbers of Indians employed to build particular stretches of line. The more useful of these are used to estimate the annual size of the Indian component of the railway construction workforce. The statements sometimes contain breakdowns of the local workforces by worker categories, gender or age. Some of these are reproduced in the ensuing tables; they are as valuable in themselves as they are for the analysis undertaken in this book. The figures presented in these sources must be taken as expressions of magnitude. Most engineers had few illusions about their ability to count, with great accuracy, the numbers of Indians under their direct or indirect charge.

Table 4 details the mean daily number of construction workers employed in building the MR in district 3, from 1 July to 31 December 1855. The line of construction from Madras to Beypore (406 miles) was divided into seventeen districts of approximately twenty-four miles each. District 3 was located mainly in north Arcot whose topography presented few obstacles and required only one substantial bridge, the fifty-six arch stone bridge across the Poini river. The bridge accounts for many of the 2161 people involved in brick-laying and masonry. Overall, the 5971 persons enumerated in Table 4, represented a mean work force of 249 per construction mile.

In July and August 1856, a total of 40,790 coolies were employed on the unfinished portions of the Madras line.[16] These encompassed districts four to seventeen, inclusive, with no work under way in district 12, preliminary work (420 coolies) in district 10 and light activity in districts 16 and 17 (1756 and 1781 coolies, respectively). No work was carried out in district 15 because the supervising engineer was ill.[17] Thus, in twelve districts (excluding districts 12 and 15), encompassing some 288 construction miles, the mean workforce per mile was 142.[18]

Table 5 details employment in districts 4–17 for the entire year, 1857. It also provides information on the total rupee expenditure on construc-

---

appeared in that article, even though I use those tables as part of the evidence for this chapter. Readers who have an interest in the estimation process, may wish to consult the earlier article and the tables therein.

[16] Selections from the Records of the Madras Government (1857), no. 44; *Report of the RR Department for 1856* (Madras, 1857), pp. 256–66.

[17] Ibid.

[18] Ibid., p. 265. The numbers employed in districts 5, 9, and 11, increased after the statistics presented in the report were compiled.

tion in each district, and the number of cubic yards of earthwork and masonry completed. The expenditure figures provide, at best, a gross indication of the wages paid in each district, since they include the costs of labour and materials. The earthwork and masonry data, especially in their variation, indicate one reason for the differential utilization of labour. However, earthwork and masonry figures do not correlate well with numbers employed since labour requirements differed considerably, depending, for example, on the kind of earth that was being moved and whether it was being cut or embanked. Overall, 47,596 were employed daily on the 336 construction miles with a mean workforce of 142 per mile. The numbers employed varied considerably, despite the districts being of similar size.

Extensive employment data is available for construction on the 416 miles of the EIR within Bengal, from Burdwan to the Kurumnassa river. These statistics, compiled in the office of the CE, George Turnbull, provide a break down of the number of workers employed daily in December 1858 (Table 6) and for an entire twelve-month period from 1859–60 (Table 7). The latter statistics, like those in Table 4, detail the job composition of the workforce. The closeness of the overall totals in Tables 6 and 7—115,635 and 118,791, respectively—suggests that all the different job categories in Table 7 are lumped together as labourers in Table 6.

The most striking characteristic of the Bengal data is the large numbers employed. Table 6 generates an overall mean of 278 workers per mile, while the mean in Table 7 is 286. The overall averages, however, must be compared with the division averages which exhibit considerable variations (ranging from 125 to 598 in December 1858, and from 138 to 568 in 1859–60). The variations, however, are one of the attractions of the data, since they represent the differences caused by particular topographical conditions, the stage of construction (e.g. preparing the road-bed by cutting or embanking required more workers than laying the permanent way), and the availability of labour. In the Monghyr division, for example, the only tunnel on the line demanded a lot of labour when it was found that the last 300 feet of its 900-foot length consisted of quartz rock which was so hard that, for a considerable period of time, only four feet per month could be tunnelled.[19] In Hallohur, the line went through an area where the Ganges annually overflowed its banks for many miles, thus much labour was needed to build bridges and embankments.[20] Sections

[19] Edward Davidson, *The Railways of India* (London, 1868), pp. 170–1.

[20] IOL&R, L/PWD/3/58, Bengal RR Letters, no. 30 dated 19 May 1859; Report of George Turnbull, CE, 18 February 1859, pp. 11–12.

of the north Rajmahal division were very unhealthy and consequently, labour was in short supply.[21]

The Bengal portion of the EIR was constructed through difficult terrain, which contributed to the largest employment per mile identified so far. This point is highlighted by the comparison between construction within Bengal, and the continuation of the EIR through the North Western Provinces, where much less labour appears to have been used. Tables 8 and 9 show construction employment in some areas of the North Western Provinces during 1855 and 1860, respectively. However, the data from the North Western Provinces is flawed by omissions and defective returns.[22] Moreover, the survival of the 1860 employment returns introduces, in the absence of the intervening years, a downward bias, since by that time the bulk of the work had been completed in many areas, while it was just getting under way in the sometimes difficult, westernmost sections of the line.[23]

Data from one mile of construction of the SP&DR between Lahore and Amritsar reinforces my suspicion that the North Western Provinces' figures err on the low side.[24] The SP&DR data generates a daily mean of 126 workers; admittedly, it relates only to one mile, but that was a mile in flat terrain and without a bridge. The same data provides an interesting day-by-day break-down of the job categories of the workers, which shows how the numbers employed varied according to the construction tasks being undertaken.

More evidence comes from western India and the construction of the GIPR. In 1856, the construction of some 141 miles utilized 31,538 persons for a mean of 224 per mile.[25] This figure excludes employment on the Western Ghat inclines, which I will discuss later. Table 10 provides employment data on various construction contracts (Ghat construction excluded) from 1858 to 1863. Two salient points emerge from the data in Table 10: first, as in the other tables, the considerable variation in the means and, second, the fact that the means are low, with the overall mean of 73 employed daily per mile being close to those for the North Western Provinces. The GIPR figures for 1863–4 were also seen as low by the Bombay Government, especially in contracts 13, 14, and 17. This was

[21] Ibid., pp. 7–8.

[22] See sources cited for table 9.

[23] Davidson, pp. 133–203, provides a good description of the various physical conditions encountered by the EIR builders.

[24] Kerr, *IESHR*, table 7.

[25] W.J. Hamilton, *Report to the Shareholders of the Great Indian Peninsula Railway Company on the Progress of the Works in India* (London, 1857), p. 27.

explained by the presence of cholera (contract 14) and the unhelpful observation that 3200 was 'a very small number' in contract 13.[26] Unremunerative rates of pay were blamed for the low numbers in contract 17, which suggests wage-responsiveness among Indian labourers. It also appears that the intensity of construction (a given stretch of line could always be built more slowly, thus using less labour per day) was low in many of the contracts listed in Table 10 (some of these are discussed in Chapter three), and/or within each contract, construction was under way only on a portion of the line.

Table 11 details construction employment on various sections of the Indian Midland Railway during the second quarter of 1886. This table again emphasizes the considerable variation in employment even in neighbouring sections of the same line. Table 12 provides some additional information from various time periods and areas which, though less-detailed, can be used to build estimates.[27] The Sind data in Table 12, though most likely tolerably correct for 1860, probably understates the longer-term employment on that line. By September 1860, when the return in question was made, the Sind Railway, Karachi to Kotri, was nearing completion; the entire line was officially opened for traffic on 13 May 1861. Indeed, other sources refer to 12,000 workers reportedly left unpaid in 1858, when the Brays abandoned their contract.[28] The 40,000 on the PNSR line in 1880, were the result of a great push to speed-up construction: 'Every possible means to hasten the work was adopted; labour was collected from down the country and at long distances, so that over 40,000 men of all classes were employed during the height of the working season.'[29] This substantial figure illustrates again how the

[26] *Annual Report of the Administration of the Bombay Presidency for the Year 1863–64*, pp. 80–2.

[27] Sawyer, *MPICE*, 97 (1889) provides excellent information on the numbers employed on the construction of the West of India Railway and harbor works, from 1884 to 1887. 1886 saw the heaviest employment averaging 14,212, in the five best working months, and 6,987 in the three worst monsoon months. Unfortunately, one cannot disentangle those employed on the Marmagao harbor works, from those building the railway.

[28] John Brunton, *John Brunton's Book* (Cambridge, 1939), pp. 98–9. At that time, 93 miles were under construction. However, 12,000 smacks too much of a conventional number, so I have not used it in my estimates.

[29] IOL&R, P/1704, PWD, RR Construction Progs., March 1881, no. 97: 'Report by Mr F.L. Diblee, late Engineer-in-Chief of the Ratial–Pindi section of the Punjab Northern State Railway, in connection with the works of construction in that section of the line'.

intensity of the construction effort, i.e. the management decision to push for rapid completion and to recruit and use large numbers of workers and a corresponding quantity of plant regardless of cost, was a significant variable in determining the size of a particular construction workforce.

The evidence presented so far ranged across the subcontinent and encompassed a variety of construction conditions.[30] Additional information would be welcome; a further intricate study of the records might provide more information which, I believe, would not materially improve estimates of the aggregate size of the construction workforce.[31] J.R.T. Hughes has suggested that in the British case, the differences in the number of workers employed to build similar lengths of railway at different locations and/or at different times, may be explained by the following: 'the technical apparatus used, the ratio between men and machines employed, the relative efficiency and inefficiency of labour, the nature of the engineering problems involved', and the intensity of the construction.[32] Some of these factors are examined in Chapters two to six. Technological innovation did occur in nineteenth-century Indian railway construction which, along with the increased and standardized familiarity of all involved in the work of construction, resulted in the more efficient use of labour. I now believe I was wrong when I wrote in 1983 in my *IESHR* article that innovation did not take place and efficiencies were not achieved. Technological innovation tended to take place in activities like the sinking of wells for bridge foundations, or for riveting, where the constricted work-setting kept the labour inputs small. Innovation, therefore, did not result in large reductions in the use of labour power, yet it did speed up the completion of work. My suspicion is that these developments—routinization and innovation—translated into more rapid railway construction in the later nineteenth-century decades, such that almost the same number of workers were employed for shorter periods.[33]

[30] Construction under famine-relief conditions represented a special case. The inclusion of the large 'weak' famine gangs would inflate the employment estimates, so I do not use such figures. The use of railway works for famine relief became a more frequent and better organized response to famine conditions, as the century wore on.

[31] And the search certainly would find, as I can testify, numerous non-numerical statements of quantity: 'large bodies of workers', 'many workers', 'large and numerous gangs', and so on—all indicating the same thing, the presence of many construction workers.

[32] J.R.T. Hughes, *Fluctuations in Trade, Industry and Finance: A Study of British Economic Development* (Oxford: Clarendon Press, 1960), p. 199.

[33] The extension of the railway network itself was a factor in expeditious construction. Workers, plant, and supplies could be more easily transported to work-sites;

Work became more intense and better organized. Something of these changes is captured in the data presented in Tables 4 to 12 as is, most surely, the need to hire more or less people, depending on the nature of the engineering (terrain) problems and/or the stage or intensity of construction. Moreover, another variable—health conditions in particular areas (a variable not listed by Hughes since its effect in Britain was less) is also covered in the data. Water shortages, the endemic presence of epidemic diseases like cholera and the debilitating effect of malaria, could keep labour in short supply, notwithstanding the recruiting efforts or the wages offered. In short, the data presented so far, despite its many shortcomings, offers a defensible basis for the estimates.

Table 13 provides a summary of the employment data. These figures provide the basis for suggesting that on an average, 126–55 workers were employed daily in India in the period 1855–1900, to construct one mile of railway; 126 is the weighted mean, i.e. the total of the daily employment column (700,177) divided by the total of the distance column (5541). The unweighted mean—the total of the mean employment daily per mile column (3245) divided by the total number of cases (21), generates the larger number of 155 per mile. Both means are vulnerable to skewing influences.

However, there is a deliberate omission in the data which forms the basis for Table 13 that deflates the employment totals in the late 1850s and early 60s, namely the absence of employment figures from the construction of the Bhore and Thal Ghat inclines, east of Bombay City up the precipitous Western Ghats. These two great works employed, within a compressed distance and a short period, a huge agglomeration of construction workers. Therefore, it is more accurate, in an estimated employment series, to present the employment totals separately for the Ghats; then to add the net employment excess (beyond 126–55 per mile) created by these extraordinary works separately into the estimates only for the years during which the Ghat inclines were under construction.

The Bhore Ghat climb of 15 miles 6 furlongs, was under construction from early 1856 to March 1863, while the 9 miles and 4 furlongs-long Thal Ghat was started in February 1858 and completed in 1865. The steep slopes and the rugged terrain of the Ghats presented formidable obstacles to the railway builders. The construction season in the Ghats was only about eight months as the monsoon rains stopped most work.

---

workers were not kept idle by the non-arrival of construction materials; workers could be paid off and transported home as soon as their task was finished. See the discussion of the Kistna bridge construction in chapters 4 and 5.

The Bhore Ghat employed some 10,000 people in 1856, rising to over 20,000 in early 1857. In 1859, some 10,000 people on an average were employed (except during a cholera outbreak in April–June when most of the workers fled). By October 1860, 18,000 people were employed. This figure rose to 30,000 in November 1860, and peaked at 42,000 in January 1861. In the next season, workmen averaged 24,000, while 33,000 were employed in January 1862. The finishing work, in the last working season, probably used less people. The Thal, for which few figures are available, appears to have used, on an average, some 10,000 workers in the seasons 1856–62; this fell to some 7000 in the final years, 1862–5.[34] Table 14 gives the yearly employment estimates for construction in the two ghats.

A daily average of 126 to 155 persons per mile, is a defensible estimate of the number of Indians employed to construct the railways of India in the period 1855 to 1900. Table 2 provides one column with 126 as the multiplier and another with 155 as the multiplier. To the aggregate figures for the years 1859–1900, generated by the application of these two estimated averages, is added the extraordinary employment in the years 1859 to 1865, created by the building of the Western Ghats inclines. Moreover, although the two great inclines represented special cases, extra large concentrations of labour did not cease with the completion of the GIPR inclines. Other substantial inclines were built. Furthermore, 113 bridges, costing not less than Rs 6,00,000 each, were built by 1909, as were 72 tunnels costing not less than 1,50,000 rupees each.[35] Table 15 lists a few of the bridges and the estimated employment required for each. Earthworkers building the training-bund and the approach bank for the Curzon bridge across the Ganges at Allahabad, numbered between 6000 and 7000 in the period from December 1903 to January 1904. A glimpse of this scene, which must have seemed like a procession of ants, from a distance, is captured in plate 1. One hundred workers per 100 feet of embanking was considered sufficient, though a few energetic petty contractors used 150, and, for a certain period, one contractor employed over 500, for one 100-foot chain.[36]

[34] See the source note to table 14.

[35] IOL&R, V/27/722/18, 'Indian Railways. Statement showing cost and particulars of some of the large railway bridges in India up to 31 December 1909 costing not less than Rs 6,00,000 each'; 'Statement showing cost and particulars of the important tunnels on the railways in India up to the 31 December 1909 costing not less than rupees 1,50,000 each'. Many bridges and tunnels cost a good deal more than these minima.

[36] Gales, *MPICE,* 174 (1908), p. 7.

Some bridge construction figures also appear in other data, presented above: for example, in the EIR data in Tables 6, 7 and 9, and in the PNSR data in Table 12. Much of the bridge and tunnel employment, however, is not represented. This fact, along with the fact that many difficult constructions were undertaken after 1865, leads me to assert that 126 persons per mile is a conservative estimate.[37] Indeed, in the estimates I published in 1983 in *IESHR*, I utilized the daily per mile average of 150 persons. I still believe 150 to be a reasonable estimate, even though it falls at the upper end of the 126–55 range suggested here. The actual employment on any particular stretch of line, of course, varied a good deal, depending on a variety of factors.

The figure of 126–55 Indians per mile (adding that of the Europeans barely nudges the mean upwards) is substantial. It is much larger, for example, than the 70 per mile that was occasionally reached in Britain between 1847 and 1860. And it was with the British experience, of course, that railway engineers in India made comparisons. Those engineers quickly came to the conclusion that two to three Indians were needed to match the work output of one British worker.[38] The 1:2–3 ratio provides corroboration for the figures suggested here, and for the assertion that they are conservative. Moreover, the estimates of miles under construction, presented in Table 1, may understate the number of miles under construction, especially for some years in the earlier decades. The combination of the estimated 126–55 Indians employed per mile, with the estimates of the miles annually under construction, produces the estimates of the annual employment aggregates.

The data presented in Table 2 concludes the task of estimation. The annual estimates of miles under construction presented in Table 1 are multiplied by the lower and higher employment per mile estimates, 126 and 155, respectively, to produce the two columns representing the annual employment of Indians in railway construction in the period 1859 to 1900. For the years when work was in progress on the GIPR Ghat inclines, the employment figures given in Table 2 include the net extra employment generated by the ghat construction (i.e. the yearly figures

[37] Among those difficult constructions were the Delhi–Amritsar line, with its great river-crossings, and, at the end of the period under study, the Bengal–Nagpur line.

[38] Berkley, *MPICE*, 19 (1859–60) p. 606; Davidson, p. 102; Burge, *MPICE*, 77 (1884), pp. 337–41. Low wages encouraged labour-intensive construction methods, and, in turn, the Indian workers had a lower per capita productivity as the result of endemic diseases (e.g., malaria), poor diets (hence less strength and smaller physiques), and the practice of employing substantial numbers of women and children. British employment figures can be obtained from Hughes, table 84 and p. 197.

presented in Table 14 minus the amount that would have been used at 126 or 155 per mile). The two columns in Table 2 indicating the possible size of the European component of the construction workforce, are based on the ratios derived earlier in this chapter. The second of these two columns represents European employment in the form of a three-year moving average, to account for the smaller year-to-year fluctuations in European employment caused by the increased length of the periods for which Europeans were hired.

The figures presented in Table 2 are approximations that indicate rough magnitudes around which considerable variations existed. Moreover, the actual employment in any particular stretch of construction varied, depending on a variety of factors: Tables 4 to 14 display some of these variations. The 126–55 per mile figures, in the absence of other information, can be applied to particular projects only with considerable caution, and, if necessary, with additions or subtractions based on the historian's knowledge of the local countryside and its social conditions. It is risky to use the per mile estimate to develop estimates of person-years of employment in particular localities for which other information is lacking.[39] Among other factors, the amount of time it took to build a mile of railway in nineteenth-century India, varied considerably. I have suggested an average of 2.5 working seasons, for reasons indicated in Chapter 2. At the long-end were the eight seasons needed for the Bhore Ghat, and, at the short-end, the 101 days used to build the 134 miles of the Kandahar State Railway, in 1878–9.[40] Also, it must be emphasized, that all miles in any given project were not under construction for the entire time it took to complete the project. For example, during the final three working seasons on the Bhore Ghat, work was concentrated on three to four miles upon which the great majority of the immense workforce was concentrated. It usually took some three to five years for a complete line, e.g. the Delhi–Amritsar, the Madras–Beypore, the EIR Jubbulpore extension and the Indus Valley lines, to be built and opened for traffic. Individual miles within a line were usually built more quickly, but normally over at least two working seasons.

It also must be emphasized that the employment estimates do not tell

[39] Sunanda Krishnamurty, 'Real Wages of Agricultural Labourers in the Bombay Deccan, 1874–1922', *IESHR*, 24:1 (January–March 1987), pp. 90–1. Regression to the mean makes the pan-Indian aggregates more accurate than an attempt to apply estimated means to particular localities.

[40] Moyle, *MPICE*, 61 (1880), p. 289. The physical conditions were such as to permit the installation of the permanent way directly on the existing terrain, with very little cutting or embanking.

us how many individual Indians found employment in railway construction during the course of a particular year. The construction workforce was volatile. Individuals were continually coming to or leaving construction work. They came to find work. They left work when they became satisfied with what they had earned, or when they were dissatisfied with the working conditions. Locally recruited labour quit when the construction sites became too distant from their villages. Many workers died, and many more became ill and had to stop working. Epidemics of disease caused workers to flee. In short, the turnover was high and many more individuals worked at railway construction than the averages presented in Table 2 suggest. It would be reasonable to suggest that the employment figures in Table 2 could be multiplied by two or three to obtain a sense of how many individuals in the course of a year were engaged in railway construction work, for longer or shorter periods. Moreover, construction labour was soon joined by reconstruction labour, as defects in the early lines became evident, as projects to double-track some lines were undertaken, and as the normal wear and tear of railway operations took their toll on the permanent way.

APPENDIX II

Tables

TABLE 1

Miles Open (1853–1900), and Miles Under Construction, 1859–1900

Year	Total route miles open	Miles opened in current year	Miles under construction
1853	20	20	
1854	71	51	
1855	169	98	
1856	272	103	
1857	287	15	
1858	427	140	
1859	625	198	1061
1860	838	213	1602
1861	1587	749	1295
1862	2333	746	998
1863	2507	174	943
1864	2958	451	831
1865	3363	405	769
1866	3563	200	545
1867	3929	366	509
1868	4008	79	803
1869	4255	247	943
1870	4771	516	856
1871	5074	303	775
1872	5369	295	1005
1873	5697	328	1008
1874	6226	529	899
1875	6541	315	776
1876	6860	319	1358

(TABLE 1: *Cont.*)

Year	Total route miles open	Miles opened in current year	Miles under construction
1877	7159	299	1324
1878	8058	899	1387
1879	8333	275	1528
1880	8995	662	1318
1881	9723	728	839
1882	9982	259	1519
1883	10198	216	1860
1884	11371	1173	1775
1885	11950	579	1710
1886	12559	609	1881
1887	13370	811	2367
1888	14135	765	2090
1889	15331	1196	1957
1890	15842	511	1512
1891	16690	848	1508
1892	17098	408	1261
1893	17774	676	1276
1894	18155	381	1403
1895	18712	557	1795
1896	19367	655	2064
1897	20228	861	2732
1898	21103	875	2962
1899	22529	1426	2369
1900	23627	1098	1592

SOURCE: Route miles open come from Morris and Dudley, 'Railway Statistics', pp. 194–5.

NOTE: Morris and Dudley remove the figures for Burma (where lines were opened from 1877 onwards). This source also provides the figures for 1901 and 1902, to complete the estimate of miles under construction through 1900.

TABLE 2

Estimated Yearly Construction Employment, 1859–1900

Year	Miles under construction	Indian employment (126 per mile)	Indian employment (155 per mile)	European employment	European employment (3-year moving average)
1859	1061	155536	185580	819	
1860	1602	238702	284435	641	659
1861	1295	200020	236850	518	519
1862	998	157598	185815	399	431
1863	943	132558	159615	377	369
1864	831	110446	134255	332	339
1865	769	100634	122645	308	286
1866	545	68670	84475	218	243
1867	509	64134	78895	204	248
1868	803	101178	124465	321	301
1869	943	118818	146165	377	347
1870	856	107856	132680	342	326
1871	775	97650	120125	258	312
1872	1005	126630	155775	335	310
1873	1008	127008	156240	336	324
1874	899	113274	139345	300	298
1875	776	97776	120280	259	337
1876	1358	171108	210490	453	384
1877	1324	166824	205220	441	452
1878	1387	174762	214985	462	471

(TABLE 2: *Cont.*)

Year	Miles under construction	Indian employment (126 per mile)	Indian employment (155 per mile)	European employment	European employment (3-year moving average)
1879	1528	192528	236840	509	470
1880	1318	166068	204290	439	386
1881	839	105714	130045	210	343
1882	1519	191394	235445	380	352
1883	1860	234360	288300	465	410
1884	1775	223650	275125	444	446
1885	1710	215460	265050	428	447
1886	1881	237006	291555	470	497
1887	2367	298242	366885	592	528
1888	2090	263340	323950	523	535
1889	1957	246582	303335	489	463
1890	1512	190512	234360	378	415
1891	1508	190008	233740	377	357
1892	1261	158886	195455	315	337
1893	1276	160776	197780	319	328
1894	1403	176778	217465	351	373
1895	1795	226170	278225	449	439
1896	2064	260064	319920	516	549
1897	2732	344232	423460	683	647
1898	2962	373212	459110	741	672
1899	2369	298494	367195	592	577
1900	1592	200592	246760	398	

TABLE 3

European Construction Employment, 1850–9

Year	RAILWAYS									
	EIR	GIPR	MR	Sind	Punjab	BB&CIR	CSER	EBR	GSIR	Total
1850	6	11								17
1851	12	11								23
1852	29	23								52
1853	67	31	11			14				123
1854	130	41	25			14				210
1855	183	51	57	7		28				326
1856	247	80	84	6	7	34				458
1857	297	117	107	2		50				573
1858	297	153	118	25	32	78		13		716
1859	265	212	123	3	45	124	12	22	13	819

SOURCE: *Railway Report, 1859*, pp. 10–30.

NOTE: The composite table on page 30 of the *Report* clearly includes operating line employees, especially where the EIR is concerned, so I have used, where available, the figures given earlier in the *Report*, which appear to focus on the construction staff. Even so, the 1857–9 EIR figures may be high. The mutiny may account for the blank for 1857 for the Punjab portion of the SP&DR. The jump to 25 in 1858 for the Sind portion of the SP&DR may be an error of some sort, but it may also include the contractor's staff. If that is so, then the 1859 figure of 3 is much too low, since the Sind line was still under construction in 1859, and, moreover, the contractor had failed, and construction was proceeding departmentally. Subsequent annual reports do not provide separate information on Europeans employed in construction.

TABLE 4

Work People Employed Daily on the Madras Railway, District 3, for the Half-Year ending 31 December 1855

	Maistries & conicopolies	Coolies	Women	Boys	Carpenters	Smiths	Total
Earthwork & ballasting	9	1413	1986	114	1	4	3527
Bricklaying & masonry	15	489	903	690	59	5	2161
Permanent way	15	187	32	34	15	0	283
Total	39	2089	2921	838	75	9	5971

SOURCE: Selections From the Records of the Madras Government (1857), no. 54: *Report of the Railway Department for 1856*, p. 253.

NOTE: Maistries were foremen and conicopolies were clerks.

TABLE 5

Mean Daily Employment, the Total Rupee Expenditure, and the Amount of Earthwork and Masonry Completed, in Each District, Madras Railway, 1857

District	Employment	Rupee expenditure	Earthwork (cubic yards)	Masonry (cubic yards)
4	5,306	4,13,082	5,26,172	12,106
5	4,451	2,39,019	5,08,469	14,050
6	5,032	2,61,804	9,37,665	3,829
7	3,580	99,519	4,50,572	9,818
8	1,698	99,519	1,31,501	6,413
9	3,548	1,58,616	6,25,785	6,086
10	1,906	78,222	5,16,825	120
11	5,328	1,77,447	4,13,952	4,679
12	45	36,556	3,59,083	0
13	3,114	1,54,344	5,73,805	15,039
14	7,369	2,75,632	10,42,060	4,013
15	2,390	98,811	4,39,262	433
16	2,610	1,43,071	6,99,046	8,949
17	1,219	1,40,989	3,31,154	4,504
Total	47,596	23,76,631	75,55,354	90,039

SOURCE: Selections from the Records of the Madras Government, no. 53a; *Report of the RR Department for 1857*, pp. xxii–xxix.

NOTE: Each district was approximately twenty-four miles in length. The rupee expenditures for districts 7 and 8 are as given in the *Report*; one of them is, most likely, a misprint in the original.

TABLE 6

Mean Daily Employment in December 1858, on Construction Divisions of the East Indian Railway, Bengal

Division	Length (miles)	Average no. of labourers per day	·Average per mile
South Birbhum	45	13,616	303
North Birbhum	32.5	9,419	290
South Rajmahal	23.5	6,035	257
Central Rajmahal	26	7,723	297
North Rajmahal	25	7,346	294
Pirpainti	26	10,815	416
Bhagalpur	23	5,931	257
Jahangira	20	2,504	125
Monghyr	31.5	13,821	439
Kiul	7.5	2,244	299
Hallohur	14	8,372	598
Barh	31.5	9,636	306
Patna	31.5	7,561	240
Soane and		8,997	
Soane bridge	78.5	1,615	135
Total	415.5	115,635	278

SOURCE: IOL&R, L/PWD/3/58, Bengal RR Letters, no. 30, 19 May 1859. Report of George Turnbull, CE, 18 February 1859, p. 20.

TABLE 7

Mean Number of Workers Employed Daily from 31 May 1859 to 31 May 1860, on Construction Divisions of the East Indian Railway, Bengal

Division	Excavators	Brickmakers	Bricklayers	Labourers	Carpenters	Sawyers	Blacksmiths	Total	Average per mile
South Birbhum	755	376	312	9,070	417	69	498	11,497	255
North Birbhum	3,185	452	306	4,460	118	67	14	8,602	265
South Rajmahal	4,773	153	65	5,275	94	25	22	10,407	443
Central Rajmahal	4,387	none	260	4,508	327	101	89	13,672	526
North Rajmahal	4,853	286	483	6,903	334	18	81	12,958	518
Pirpainti	3,670	655	180	5,201	115	47	30	9,898	381
Bhagalpur	482	262	190	5,941	70	15	24	4,984	217
Jahangira	2,121	403	177	2,014	26	17	13	4,771	239
Monghyr	1,322	485	311	4,964	285	60	324	7,751	246
Kiul	663	266	106	2,763	142	39	60	4,039	539
Hallohur	2,122	903	276	4,336	137	92	84	7,950	568
Barh	2,144	920	235	3,672	122	53	13	7,159	227

(TABLE 7: *Cont.*)

Division	Excavators	Brickmakers	Bricklayers	Labourers	Carpenters	Sawyers	Blacksmiths	Total	Average per mile
Patna	905	103	125	3,045	122	15	26	4,341	138
Soane	2,862	874	548	4,336	84	35	33	8,772	137
Soane bridge	72	69	1,494	204	38	81	32	1,990	
Total	38,316	6,207	5,068	64,692	2,431	734	1,343	118,791	286

SOURCE: Statement from George Turnbull's office reproduced in 'Rajmahal, Its Railway and Historical Association', *Calcutta Review*, vol. 36 (March 1861), p. 11.

NOTE: Row totals for Central Rajmahal and Bhagalpur and the column totals of excavators and labourers do not correspond to the sum of the numbers in those rows and columns. Central Rajmahal, for example, totals 9672. However, since it was the site of the laborious Sitapahar-cutting, the total, as given, 13,672, is plausible and the error is probably a mistake in the excavator or labourer cell. Moreover, the overall row and column totals agree at 118,791—a number for which there is other corroboration. I have, therefore, not adjusted the individual row and column totals except for Soane bridge which is clearly 1990 rather than 4990, as stated in the original table. Table 6 divisional lengths have been used to provide the average per mile figure in this table.

TABLE 8

Mean Daily Employment, EIR in the North Western Provinces, June 1855

District	Distance	Excavators and labourers	Carpenters	Bricklayers	Smiths
Mirzapur	95	8,487	158	340	54
Fatehpur	121	3,700	10	1	7
Benares	57	1,393	—	—	—

SOURCE: IOL&R, L/PWD/3/51,Bengal RR Letters, collection 3 to letter no. 2 of 1856.

TABLE 9

Mean Daily Employment, EIR in the North Western Provinces (January, April and June 1860)

District	Distance	Excavators and labourers	Others	Total daily mean	Daily mean per mile
Mirzapur east	28	1,797	716	2,513	90
Middle Mirzapur	32	2,525	1789	4,314	135
Mirzapur west	31.5	2,022	68	2,090	67
Tons bridge	1	800	156	956	956
Jumna bridge (Allahabad)	1	1,436	631	2,067	2,067
Cawnpore	27	555	165	720	27
Agra	24	52	39	91	4
Agra branch	13	435	43	478	37
Jumna bridge (Delhi)	1	318	125	443	443
Hindan bridge	1	73	100	173	173
Etawah	44	2,325	58	2,383	54
Total	286	20,186	4,085	24,271	85

SOURCE: IOL&R, North Western Provinces, RR Dept. Progs: no. 98A, 23 April 1860; no. 75, 25 June 1860; no. 64, 18 September 1860.

NOTE: 'Others' includes carpenters, bricklayers and masons, smiths, peshrajas, vicemen, well-sinkers, quarrymen, stone dressers and brick moulders. I use one mile as the distance for all the bridges, though only one bridge exceeded that distance and the others were less than a mile. However, bridge approaches covered some distance and they required additional labour. A more detailed version providing the January, April and June breakdowns can be found in Kerr, *IESHR*, Table 6.

TABLE 10

Mean Daily Employment on the GIPR Construction Contracts, 1858–64

Year	Contract number	Distance	Mean daily employment	Mean daily employment per mile
1858–9	10	25	9,300	372
	12	190	10,700	56
1859–60	12	190	15,928	84
1860–1	12	190	17,538	92
	13	134.5	8,387	62
	14	109	2,000	18
1861–2	12	105	7,743	74
	13	134.5	13,435	100
	14	109	7,203	66
	15	139	11,960	86
	16	80	5,373	67
	17	119	11,255	95
1863–4	13	105	3,200	30
	14	109	3,148	29
	15	139	8,652	62
	16	80	10,390	130
	17	119	4,484	38

SOURCE: GIPR, *Annual Report of the Company for the Year 1858–59, 1859–60, 1860–61, 1861–62, 1863–64.* The only place I have found these reports is in IOL&R, L/PWD/2/107–9 RR Home Correspondence—C: Letters to and from Railway Companies, etc. Register III: GIPR Company 1858–79, enclosures to Letters.

NOTE: The reduced distances on some of the contracts in later years were the result of lines being partially completed.

TABLE 11

Construction Employment, Indian Midland Railway, April–June 1886

Area	Distance	Mean daily employment	Mean daily employment per mile
1. Jhansi–Manikpur section			
a. Karwi division	60	1,237	21
b. Banda division	56	2,182	39
c. Mhow division	65	3,033	47
2. Bhopal–Jhansi section			
a. Bhilsa division, Bhilsa to Basoda only	24	625	26
b. Etawah division	66	1,881	29
c. Lalitpur division	56	8,043	144
Totals	327	17,001	52

SOURCE: IOL&R, P/2751, PWD, RR Construction Progs., September 1886, no. 232: 'Narrative Progress Report for the quarter ending 30th June 1886', by A.C. Cregeen, Agent and Chief Engineer, Indian Midland Railway Company, 22 July 1886.

NOTE: The Karwi division had a daily average of some 2200, but the contractor ran into difficulties and the number dwindled to 1237. The figures for Lalitpur division are plausible; work was only recently under way and was at the stage where cutting and embanking constituted much of it. 13,000,000 cubic feet of earthwork had already been performed, which can be compared to the 3,000,000 cubic feet done in the Etawah division by 30 June 1886. These earthwork statistics explain the substantial differences in the Etawah and Lalitpur employment figures.

TABLE 12

Miscellaneous Construction Employment Figures

Line	Date	Numbers employed daily	Distance	Mean daily employment per mile
Bombay, Baroda and Central India	early 1857	11,600	93	125
Bombay, Baroda and Central India	May 1860	26,098	277	94
Sind (section of the SP&DR)	1860	7,635	109	70
Punjab Northern State Railway Rawalpindi division	1879–80	40,000	73	548
Sind–Pishin State Railway Three upper divisions	June-August 1886	18,709	124	151
Assam–Bengal Railway, section II	1896	10,000	41	244

SOURCE: Berkley, *An Address*, p. 11; MSA, Bombay PWD (Railways), 1860, vol. 56: compilation 472; IOL&R, P/1704, PWD, RR Construction Progs., March 1881, no. 97; IOL&R, P/3001, PWD, RR Construction Progs., January 1887, no. 248; IOL&R, PWD, RR Construction Progs., July 1896, no. 234.

TABLE 13

Statements of Mean Daily Employment per Construction Mile

Location	Year	Distance	Daily employ-ment	Mean daily employment per mile
Madras Railway	1855	24	5,971	249
,,	1856	288	40,790	142
,,	1857	336	47,596	142
Bombay, Baroda and Central India	1857	93	11,600	125
,,	1860	277	26,098	94
East Indian Railway, Bengal	1858	416	1,15,635	278
,,	1859–60	416	1,18,791	286
East Indian Railway, NWP	1855	273	14,150	52
,,	1860	524	43,841	84
Punjab Railway	1860	1	126	126
Sind Railway	1860	109	7,635	70
Great Indian Peninsula Railway	1856	141	31,538	224
,,	1858–9	215	20,000	93
,,	1859–60	190	15,928	84
,,	1860–1	434	27,925	64
,,	1861–2	687	56,969	83
,,	1862–3	552	29,874	54
Punjab Northern	1879–80	73	40,000	548
Indian Midland	1886	327	17,001	52
Sind–Pishin	1886	124	18,709	151
Assam–Bengal	1896	41	10,000	244
Total		5541	700,177	3245

SOURCE: Tables 4 to 12 and the text.

NOTE: The weighted mean, 700,177, divided by 5541, is 126; the unweighted mean, 3245, divided by 21, is 155.

TABLE 14

Mean Daily Employment on the Bhore and Thal Ghat Inclines, 1856–65

Year	Mileage	Estimate of employment
1856	16	10,000
1857	16	20,000
1858	25	30,000
1859	25	25,000
1860	25	40,000
1861	25	40,000
1862	25	35,000
1863	10	15,000
1864	10	7,000
1865	10	5,000

SOURCE: MSA PWD, Railway, 1863, vol. 10, compilation 211, Opening of the Bhore Ghat Incline. This collection includes a pamphlet, *Narrative of the Operations in the Bhore Ghat Incline from their Commencement in 1850 to their Completion in March 1863* (Bombay: Education Society's Press, 1863), and GIPR, *Annual Report of the Company for the year 1858–59, 1860–61, 1861–62* and *1863–64*. Mileage figures have been rounded off. Also useful is Berkley, *MPICE*, 19 (1859–60), pp. 594–8.

TABLE 15

Estimated Employment at Selected Great Bridges

Bridge	River	Date under construction	Employment
Empress	Sutlej	1873–1878	5000 plus
Dufferin	Ganges	1881–1886	7000
Betwa and nearby bridges	Betwa	1885	40,000
Sher Shah	Chenab	1888–1890	5000
Bezwada	Kistna	1890–1893	3000–6000

SOURCES: Bell, *MPICE*, 65 (1881), p. 257; Walton, *MPICE*, 101 (1890), p. 24; Spring, Technical Paper no. 71. The figures presented by the authors appear to represent peak daily maxima. The Betwa information can be found in IOL&R, PWD, RR Construction Progs., March 1885, and represents the peak that was achieved in February 1885.

Bibliography

GOVERNMENT RECORDS, UNPUBLISHED

1. *Government of India Records held in the India Office Library and Records, London*

   Collections (Enclosures) to Railway Letters from Bombay, 1846–79

   Collections (Enclosures) to Railway Letters from Madras, 1839–79

   Compilations and Miscellaneous. 'Public Works Old Series' (Collections)

   Public Works Proceedings

   Railway Letters and Enclosures from Bengal and India, 1845–79

2. *Provincial Records held in the India Office Library and Records*

   North Western Provinces, Public Works Proceedings

3. *Government of India Records held in the National Archives of India, New Delhi*

   Foreign Department, Political Proceedings
   Home Department, Public Proceedings
   Public Works Department, Proceedings
   Railway Department, Proceedings

4. *Bombay Presidency Records held in the Maharashtrian State Archives, Bombay*

   Public Works Department (Railway), Compilations

5. *Madras Presidency Records held in the Tamil Nadu State Archives, Madras*

   Public Works Department (Railway) Proceedings

GOVERNMENT RECORDS, PUBLISHED

Bombay Government. *Annual Report of the Administration of the Bombay Presidency for the Year 1863–1864.*

Census of India, 1871, 1881

Davies, R.H., R.E. Egerton, R. Temple and J.H. Morris. *Report on the Revised Settlement of the Lahore District in the Lahore Division.* Lahore: 1860.

Government of India, Director of Railway Construction. *Technical Papers.*

Government of India, Indus Valley State Railway. *Administration Report of the Construction of the Railway from its Commencement to the Close of the Financial Year 1878–79.* Roorkee: Thomason Civil Engineering College Press, 1879.

Government of India, Railway Department. *Catalogue of Technical Papers* issued by the Technical Section of the Railway Board. Delhi: Government of India Press, 1925.

Government of India, Railway Board. *Technical Papers.*

History of Services of the Officers of the Engineer, Accounts, and State Railway Revenue Establishments, including the Military Works Department under the Military Department, 4th edn, corrected to 31 December 1888. Calcutta: Superintendent of Government Printing India, 1889.

Madras Government. Madras Railway Reports, 1853 to 1855.

Madras Government. *Selections from the Records of the Madras Government.*

Reports on the Nerbudda Bridge by Colonel J.S. Trevor, C. Curry and F. Mathew (1868).

Trevor, Major-General J.S. *Administration Report on Indian State Railways From Their Commencement to the End of 1879–80.* Calcutta: Superintendent of Government Printing, 1881.

PARLIAMENTARY PAPERS

Reports to the Secretary of State for India in Council on Railways in India: Command number (session)

c. 2669 (1860); *c.* 2826 (1861); *c.* 3009 (1862); *c.* 3168 (1863); *c.* 3354 (1864); *c.* 3521 (1865); *c.* 3696 (1866); *c.* 3856 (1867); *c.* 4035 (1867–8); *c.* 4190 (1868–9); *c.* 163 (1870); *c.* 418 (1871); *c.* 643 (1872); *c.* 838 (1873); *c.* 1070 (1874); *c.* 1369 (1875); *c.* 1584 (1876); *c.* 1823 (1877); *c.* 2179 (1878); *c.* 2386 (1878–9); *c.* 2683 (1880); *c.* 2999 (1881); *c.* 3328 (1882); *c.* 3692 I. (1883); *c.* 4080 (1884); *c.* 1490 (1884–5); *c.* 4832 (1886); *c.* 5122 (1887); *c.* 5444 (1888); *c.* 5770 (1889); *c.* 6089 (1890); *c.* 6409 (1890–1); *c.* 6793 (1892); *c.* 7067 (1893–4); *c.* 7453 (1894); *c.* 7845 (1895); *c.* 8136 (1896); *c.* 8518 (1897); *c.* 8921 (1898); *c.* 9369 (1899); *c.* 232 (1900).

Report from the Select Committee on East India (Railways). *c.* 416 (1857–8)

Copies of all contracts and agreements entered into by the East India Company or by the present Council for India, with any company formed for making Railways, Public Roads, Canals, Works for Irrigation, or other Public Works in India. *c.* 259 (1859).

Report from the Select Committee on East India Finance. *c.* 327 (1872) and *c.* 354 (1873).

Report from the Select Committee on East India (Public Works). *c.* 333 (1878).

Report from the Select Committee on East India Railway Communication. *c.* 284 (1884).

MANUSCRIPT COLLECTIONS

1. *In the India Office Library and Records, London*
   Carrington Papers. Eur. Mss B212
   Fowler Letters. Eur. Mss C 401
   Going Typescript. Eur. Mss C 378
   Mann Letters. Photo. Eur. 197
   Rayne Papers. Eur. Mss D.904
   West Collection. Eur. Mss D.1184

2. *In the Scottish Record Office, Edinburgh*
   Dalhousie Muniments

3. *In the Archives of the Institution of Civil Engineers, London*

   Original submissions to *MPICE* for consideration for publication. These are the items referred to in my source notes as ICE, Mss. There are over 3000 of these in the period covered in this study. Many were never published. Some were published in modified form.

4. *Hertford Record Office, Hertford, England*
   Leake Papers

NEWSPAPERS

Bombay Times Overland Summary
Civil and Military Gazette
Indian Public Opinion and Punjab Times
Lahore Chronicle
Madras Times
Times of India

ENGINEERING JOURNALS, RAILWAY NEWSPAPERS & ENGINEERING SERIAL PUBLICATIONS

Engineering
Herapath's Railway Journal
Institute of Mechanical Engineers, *Proceedings*
Professional Papers on Indian Engineering
Professional Papers of the Corps of Royal Engineers, Royal Engineers Institute. *Occasional Papers*
Railway Gazette
The Bombay Builder
The Engineer
The Engineer's Journal

MINUTES OF PROCEEDINGS OF THE INSTITUTION OF CIVIL ENGINEERS (MPICE)

Anderson, Francis Philips. 'The Effects of the Earthquake in 1897 on the Shaistaganj Division of the Assam–Bengal Railway', 141 (1900), pp. 258–61.

Bamford, Charles Frederic. 'The Pahartali Locomotive and Carriage Works, Assam–Bengal Railway', 135 (1899), pp. 212–23.

Bayley, Victor. 'The Khyber Railway', 222 (1927), pp. 1–78

Beckett, William T.C. 'The Bridges over the Orissa Rivers on the East Coast Extension of the Bengal–Nagpur Railway', 145 (1901), pp. 268–91.

Bell, James Richard. 'The First Section of the Kandahar Railway', 61 (1880), pp. 274–85.

Bell, James Richard. 'The Empress Bridge over the Sutlej', 65 (1881), pp. 242–58.

Berkley, James John. 'On Indian Railways: With a Description of the Great Indian Peninsula Railway', 19 (1860), pp. 586–624.

Burge, Charles Ormsby. 'The Comparative Value of Labour in Different Countries', 77 (1884), pp. 337–41.

Clarke, Colonel Sir Andrew. 'Note on the Kandahar Railway', 61 (1880), p. 273.

Cole, Charles John. 'The Tunnels on the Second Division of the Mushkaf-Bolan Railway, India', 128 (1897), pp. 265–77.

Evans, Frank Dudley. 'Engineering Operations for the Prevention of Malaria', 200 (1915), pp. 2–61

Eves, Graves William. 'Well-Sinking on the Koyatha: Bridge, Bengal–Nagpur Railway'. 145 (1901), pp. 292–7.

Farrell, Richard Craig. 'Railway Flood-Works in the Punjab and Sind, Relative to the NorthWestern State Railway', 140 (1900), pp. 130–42.

Gales, Robert Richards. 'The Curzon Bridge at Allahabad', 174 (1908), pp. 1–40

Gales, Sir Robert Richard. 'The Hardinge Bridge over the Lower Ganges at Sara', 205 (1920), pp. 18–99.

Hargrave, H.J.B. 'Railway Bridge over the Nerbudda River', 94 (1888), pp. 347–8.

Hearn, Colonel Gordon Risley. 'The Survey and Construction of the Khyber Railway', 222 (1927), pp. 23–78.

Hearn, Gordon Risley. 'Railway Surveying on the Pipli Ghat', 14 (1902), pp. 263–75.

Johnson, Francis Robert. 'The Survey of the Manmad-Dhulia Railway India', 115 (1894), pp. 343–51.

Jones, William Arthur. 'The Tunnels on the First Division of the Mushkaf-Bolan Railway, India', 128 (1897) pp. 257–64.

Ker, Thomas. 'Concrete Quarters for Native Clerks, Guards and Menial Staff on Indian Railways', 101 (1890), pp. 186–8.

Lacey, Joseph Melville.'Floods in Southern India', 171 (1908), pp. 360–70.

M'Master, Bryce.'On the Permanent Way of the Madras Railway', 18 (1858–9), pp. 417–44.

Martin-Leake, Stephen. 'The Rupnarayan Bridge, Bengal–Nagpur Railway', 151 (1903), pp. 251–300.

Moyle, George. 'The Platelaying of the Jacobabad or Broad Gauge Section of the Kandahar Railway', 61 (1880), pp. 286–94.

Nolan, Thomas Richard. 'Slips and Washouts on the Hill Section of the Assam–Bengal Railway', 218 (1925), pp. 2–69.

Nolan, Thomas Richard. 'The Construction of the Ninth Division (Hills section) of the Assam–Bengal Railway', 177 (1909), pp. 316–44

Ramsay, James. 'The Mushkaf-Bolan Railway, Baluchistan, India', 128 (1897), pp. 232–56

Robertson, Frederick Ewart. 'The Lansdowne Bridge over the Indus at Sukkur', 103 (1891), pp. 123–200.

Robertson, Frederick Ewart. 'The Jubilee Bridge over the Hooghly', 123 (1896), pp. 406–9.

Royal-Dawson, Frederick George. 'The Indian Railway Gauge Problem', 213 (1922), pp. 15–122.

Sawyer, Ernest Edward. 'West of India Portuguese Railway and Harbour Works', 97 (1889), pp. 302–34.
Shaw, William Robert. 'The Teesta Bridge', 150 (1902), pp. 361–75.
Stoney, Edward Waller. 'Extraordinary Floods in Southern India: the causes and destructive effects on Railway Works', 134 (1898), pp. 66–118.
Stoney, Edward Walter. 'The New Chittravati Bridge', 103 (1891), pp. 135–200.
Thomas, Andrew. 'Some Bridge Failures During a Heavy Cyclone in 1905 in Gujerat, India', 166 (1906), pp. 224–8.
Thompson, Harry James. 'The New Papaghni Bridge on the Madras Railway', 122 (1895), pp. 187–92.
Thornton, William Thomas. 'The Relative Advantages of the 5 ft 6 in. Gauge and of the Metre Gauge for the State Railways of India, and particularly for those of the Punjab', 35 (1873), pp. 214–535.
Tufnell, Carleton Fowell. 'Economical River-Training in India', 72 (1883), pp. 177–84.
Upcott, Sir Frederick Robert. 'The Railway Gauges of India', 164 (1906), pp. 196–327.
Walton, Frederick Thomas Granville. 'The Construction of the Dufferin Bridge over the Ganges at Benares', 101 (1890), pp. 13–24.
Waring, Francis John. 'Indian Railways. The Broad and Narrow-Gauge Systems Contrasted', 97 (1889), pp. 106–94.
Weightman, Walter James. 'The Nilgiri Mountain Railway', 145 (1901), pp. 1–43.
Weightman, Walter James. 'The Khojak Rope-Inclines', 112 (1893), pp. 310–20.
White, Colin Robert. 'Pamban Viaduct, South Indian Railway', 199 (1915), pp. 377–87.

MANUALS AND LECTURES

Addis, A.W.C. *Practical Hints to Young Engineers Employed on Indian Railways.* London: E & F.N. Spon, 1910.
Berkley, James J. Paper on the Thul Ghaut Railway Incline: Read at the Bombay Mechanics' Institution, in the Town Hall, on Monday, 10 December 1860. Bombay: Education Society's Press, Byculla, 1861.
Berkley, James J. An Address Delivered at the Annual Meeting of the Bombay Mechanic's Institution, in the Town Hall on Saturday, 11 April 1857. Bombay: The 'Bombay Gazette' Press, 1857.

Finney, S. *Railway Construction in Bengal.* Three lectures delivered at the Sibpur Engineering College in January and February 1896. Calcutta: Bengal Secretariat Press, 1896.

Finney, S. *Railway Management in Bengal.* Three lectures delivered at the Sibpur Engineering College in February and March 1896. Calcutta: Bengal Secretariat Press, 1896.

George, E. Monson. *Railways in India.* Their economical construction and working. London: Effingham Wilson, 1894.

Joyce, H.W. *Economy in Bridge Design and Construction.* A series of six lectures delivered to the students of the Sibpur Eengineering College during March 1914. Calcutta: Bengal Secretariat Book Depot, 1915.

Joyce, H.W. Five Lectures on Indian Railway Construction and one Lecture on Management and Control. A series of six lectures delivered to students of the Sibpur Engineering College during March 1905. Calcutta: The Bengal Secretariat Book Depot, 1905.

Medley, Lt. Colonel J.G., compiler. *The Roorkee Treatise on Civil Engineering In India.* 3rd edn, edited by Major A.M. Lang. 2 vols. Roorkee: Thomason College Press, 1877.

Paper on the Bhor Ghaut Incline of the Great Indian Peninsula Railway: Read at the Bombay Mechanics' Institution, in the Town Hall, Bombay, 21 December 1857 by the President, James J. Berkley, with an Appendix by A.A. West. Bombay: Education Society's Press, 1863.

Spring, Francis J.E. *River Training and Control.* Being a description of the theory and practice of the modern system entitled the guide bank system, used in India for the control and guidance of great alluvial rivers. Simla: Government Central Printing Office, 1903.

(Wynne, Sir J.R.). *Notes on the Construction of Railways in India for Students.* Calcutta: Bengal Secretariat Press, 1902.

MISCELLANEOUS

Bombay. Memorandum by the Commission appointed to collect information on the subject of prices as affecting all classes of Government servants. (1863)

First and Second Annual Reports of the Great Indian Peninsula Railway Company, for the Years 1854–55—1855–56. Bombay: Education Society's Press, 1857.

Indian Railways. Statement showing cost and particulars of some of the large railway bridges in India up to the 31 December 1909 costing not less than Rs. 6,00,000 each.

Indian Railways. Statement showing cost and particulars of the important tunnels on the railways in India up to 31 December 1909, costing not less than Rs. 1,50,000 each.

Panjab Notes and Queries. vols 1 & 2. 1884

The Indian Economist and *The Statistical Reporter*

Report to the Shareholders of the Great Indian Peninsula Railway Company on the Progress of the Works in India. London: W. Lewis & Son, 1857.

BOOKS, ARTICLES, PAMPHLETS AND THESES

'Eastern Bengal and Its Railways', *Calcutta Review*, no. LXXI (March 1861), pp. 158–84.

Gazetteer of the Bombay Presidency. vol. 17, part 1: Poona. Bombay: Government Central Press, 1885.

Imperial Gazetteer of India. new edn, 26 vols. Oxford: Clarendon Press, 1909.

India: Views on the Bhore Ghaut, Shewing Some Railway Cuttings. Photographed by Captain C. Scott. London: J. Hogarth, 1860.

'Indian Railways', *Cornhill Magazine,* XX (July–December 1869), pp. 68–80.

Letters Indicating Practical Means for the Extensive Development of Cheap Railways in India. Calcutta: 1849.

'Opening of the Meerut and Umballa Section of the Delhi Railway, on the 14 November 1868, by His Excellency The Viceroy, the Right Hon. Sir John Lawrence.' London: 1869.

'Our Railways', *Calcutta Review*, no. LXXII (June 1861), pp. 390–405.

Professional Papers of the Corps of Royal Engineers. Index 1837–92. Chatham: The Royal Engineers Institute, 1893.

'Railway Fuel in the Punjab', *Calcutta Review*, no. XCII (1868), pp. 262–327.

'Railways in Western India', *Bombay Quarterly Review*, I (January–April 1855), pp. 281–322.

'Rajmahal, Its Railways and Historical Associations', *Calcutta Review*, no. LXXI (March 1861), pp. 110–43.

'The East Indian Railway', *Calcutta Review*, no. LXI (September 1858), pp. 230–46.

'The Sonthal Rebellion', *Calcutta Review*, no. LI (March 1856), pp. 223–64.

Adas, Michael. *Machines as the Measure off Men. Science, Technology, and Ideologies of Western Dominance.* Ithaca: Cornell University Press, 1989.

Alatas, Syed Hussein. *The Myth of the Lazy Native.* London: Frank Cass, 1977.

(Andrew, W.P.) *A Letter to the Shareholders of the East Indian and Great Western of Bengal Railways, on their Present Position and Future Prospects.* By one of themselves. London: Smith, Elder & Co., 1847.

Andrew, W.P. *Railways in Bengal* (Being the substance of a report addressed to the Chairman of the East India Company in 1849) with introductory remarks by the Editor of the *Artizan* (reprinted from the *Artizan* of June 1851), a map and appendix. London: Allen, 1853.

Antia, K.F. *Railway Track.* Design, Construction, Maintenance, and Renewal of Permanent Way with Notes on Signalling and Bridge Maintenance. Bombay: New Book Company, 1945.

Appleby, Leighton L. 'Social Change and Railways in North India, *c.* 1845–1914'. Ph.D. dissertation, University of Sydney , 1990.

Armytage, W.H.G. *A Social History of Engineering.* 3rd edn, London: Faber and Faber, 1970.

Arnold, David. 'European Orphans and Vagrants in India in the Nineteenth Century', *Journal of Imperial and Commonwealth History,* VII:2 (January 1979), pp. 104–27.

Arnold, David. 'White Colonization and Labour in Nineteenth-Century India', *Journal of Imperial and Commonwealth History,* XI:2 (January 1983), pp. 133–58.

Arnold, David. 'Medical Priorities and Practice in Nineteenth-Century British India', *South Asia Research,* 5:2 (November 1985), pp. 167–83.

Ayrton, Frederick. *Some Considerations on the Means of Introducing Railways into India, etc.*, n.d. but *c.* 1848.

Baden Powell, B. H. *Handbook of the Manufactures and Arts of the Punjab, with a Combined Glossary and Index of Vernacular Trades and Technical Terms.* Lahore: Punjab Printing Company, 1872.

Bagwell, Philip S. *The Transport Revolution from 1770.* London: B.T. Batsford Ltd., 1974.

Balandier, Georges. *The Sociology of Black Africa.* London: Deutsch, 1970.

Banaji, Jairus. 'For a theory of colonial modes of production', EPW (23 December 1972), 2498–2502.

Banaji, Jairus. 'Deccan Districts in the Late Nineteenth Century', EPW (August 1977), 1375–1404.

Banerji, A. K. *Aspects of Indo-British Economic Relations 1858–98.* Bombay: Oxford University Press, 1982.

Barber, C.G. *History of the Cauvery-Mettur Project.* Madras: Superintendent Government Press, 1940.

Bates, Crispin N. 'Regional Dependence and Rural Development in Central India: The Pivotal Role of Migrant Labour', MAS, 19:3 (February 1985), pp. 573–92.

Bawa, Vasant Kumar. 'Salar Jang and the Nizam's State Railway 1860–1883', *IESHR*, II:4 (October 1965), pp. 307–40.

Bayley, Victor. *Nine-fifteen from Victoria.* London: Robert Hale, 1937.

Bayley, Victor. *Indian Artifex.* London: Robert Hale, 1939.

Bayley, Victor. *Permanent Way through the Khyber.* London: Jarrolds, 1934.

Bayly, C.A. *Rulers, Townsmen and Bazaars: North Indian Society in the Age of British Expansion, 1770–1870.* Cambridge: Cambridge University Press, 1983.

Bayly, C.A. *The New Cambridge History of India.* vol. 2:1. *Indian Society and the Making of the British Empire.* Cambridge: Cambridge University Press, 1988.

Bell, Horace. 'Indian Railways', *Asiatic Quarterly Review,* III (January–April 1887), pp. 331–55.

Bell, Horace. *Railway Policy in India.* London: Rivington, Percival and Co., 1894.

Bell, Major Evans and Lieut.-Col. Frederick Tyrrell. *Public Works and the Public Service in India.* London: Trubner and Co., 1871.

Bell, S.P. *A Biographical Index of British Engineers in the 19th Century.* New York and London: Garland Publishing Inc., 1975.

Berg, Maxine, Pat Hudson and Michael Sonenscher (eds). *Manufacture in Town and Country before the Factory.* Cambridge: Cambridge University Press, 1983.

Berridge, P.S.A. 'The Story of the Khojak—The Longest Railway Tunnel in India', *Indian State Railway Magazine,* VII:8 (May 1934), pp. 412–16.

Berridge, P.S.A. *Couplings to the Khyber: The Story of the North Western Railway.* New York: Augustus M. Kelley, 1969.

Berridge, P.S.A. 'The Sind Peshin and Khandahar State Railway', *Indian State Railways Magazine,* IX:9 (June 1936), pp. 587–98.

Bhattacharya, Sabyasachi. 'Presidential Address, Modern Indian History', *Indian History Congress* (1982), p. 23.

Bottomore, Tom (ed.). *A Dictionary of Marxist Thought.* Oxford: Blackwell Reference, 1983.

Bourne, John. *Indian Works and English Engineers.* A Letter to His Grace the Duke of Argyll, Secretary of State for India. London:Longmans, Green & Co., 1870.

Brassey, Earl. *Work and Wages: The Reward of Labour and the Cost of Work Founded on the Experiences of the Late Mr. Brassey, a Volume of Extracts, Revised, and Partially Rewritten.* London: Longmans, Green & Co., 1916.

Braverman, Harry. *Labor and Monopoly Capital: The Degradation of work in the Twentieth Century.* New York: Monthly Review Press, 1974.

Breman, Jan. *Labour Migration and Rural Transformation in Colonial Asia.* Amsterdam: Free University Press, 1990.

Breman, Jan. *Of Peasants, Migrants and Paupers: Rural Labour Circulation and Capitalist Production in West India.* Delhi: Oxford University Press, 1985.

Breman, Jan. *Patronage and Exploitation: Changing Agrarian Relations in South Gujarat, India.* Berkeley: University of California Press, 1974.

Breman, Jan. 'Seasonal migration and co-operative capitalism; the crushing of cane and of labour by the sugar factories of Bardoli, South Gujarat, parts 1 and 2', JPS, 6:1&2 (1978–9), pp. 41–70, 168–209.

Brooke, David. *The Railway Navvy.* Newton Abbot: David & Charles, 1983.

Brunton, John. *John Brunton's Book, being the Memories of John Brunton, Engineer, from a Manuscript in his own Hand Written for his Grandchildren and now Printed.* With an introduction by J.H. Clapham. Cambridge: Cambridge University Press, 1939.

Buchanan, Daniel Houston. *The Development of Capitalistic Enterprise in India.* 1934; reprint edn, London: Frank Cass & Co., 1966.

Buchanan, Francis. *A Journey from Madras through the Countries of Mysore, Canara and Malabar.* 3 vols. London: 1807–11.

Buchanan, R.A. 'Institutional Proliferation in the British Engineering Profession, 1847–1914', *Economic History Review,* 2nd series, 38:1 (February 1985), pp. 42–60

Burawoy, Michael. *The Politics of Production.* London: Verso, 1985.

Burawoy, Michael. *Manufacturing Consent: Changes in the Labour Process under Monopoly Capitalism.* Chicago: University of Chicago Press, 1979.

Burge, C.O. *The Adventures of a Civil Engineer: Fifty Years on Five Continents.* London: Alston Rivers Ltd., 1909.

Cautley, Colonel Sir Proby T. *Report on the Ganges Canal Works: From their Commencement until the Opening of the Canal in 1854.* 3 vols. London: Smith, Elder & Co., 1860.

Chakrabarty, Dipesh. *Rethinking Working-class History. Bengal 1890–1940.* Princeton: Princeton University Press, 1989.

Chakrabarty, Dipesh. 'Class Consciousness and the Indian Working Class: Dilemmas of Marxist Historiography', *Journal of Asian and African Studies*, 23:1–2 (1988), pp. 21–31.

Chakrabarty, Dipesh. 'On Deifying and Defying Authority: Managers and Workers in the Jute Mills of Bengal 1890–1940', *Past and Present*, 100 (August 1983), pp. 124–46.

Chakravarti, A.K. *Railways for Developing Countries* (with special emphasis on India). New Delhi: Chetana Publications, 1982.

Chandra, Bipan. *The Rise and Growth of Economic Nationalism in India: Economic Policies of Indian National Leadership, 1880–1905.* New Delhi: People's Publishing House, 1966.

Charlesworth, Neil. *British Rule and the Indian Economy 1800–1914.* London: Macmillan, 1982.

Chaudhuri, M.K. (ed.). *Trends of Socio-economic Change in India 1871–1961.* Simla: Indian Institute of Advanced Study, 1969.

Chaudhuri, Sashi Bhusan. *Civil Disturbances during the British Rule in India (1765–1857).* Calcutta: The World Press, 1953.

Cohen, R., Peter C.W. Gutkind and P. Brazier. *Peasants and Proletarians: The Struggles of Third World Workers.* New York: Monthly Review Press, 1979.

Cohen, Robin, Jean Copans and Peter C.W. Gutkind (eds). *African Labour History.* Beverly Hills: Sage Publications, 1978.

Coleman, Terry. *The Railway Navvies.* A history of the men who made the railways. Harmondsworth: Penguin Books, 1968.

Cox, Arthur A., compiler. *A Manual of the North Arcot District in the Presidency of Madras.* Madras: Government Press, 1881.

Crooke, W. *The Tribes and Castes of the North-Western Provinces and Oudh.* 4 vols. Calcutta: Office of the Superintendent of Government Printing, India, 1896.

Darukhanawala, H.D. *Parsi Lustre on Indian Soil.* vol. 1. Bombay: 1939.

Das Gupta, Ranajit. 'Factory Labour in Eastern India: Sources of Supply, 1855–1946. Some Preliminary Findings', *IESHR*, 13:3 (July–September 1976), pp. 277–329.

Das, M.N. *Studies in the Economic and Social Development of Modern India: 1848–56.* Calcutta: Firma K.L. Mukhopadhyay, 1959.

Das, Rajani Kanta. *History of Indian Labour Legislation.* Calcutta: University of Calcutta, 1941.

Datta, Kalinkar. *The Santhal Insurrection of 1855–57.* Calcutta: University of Calcutta, 1940.

Davidson, Edward. *The Railways of India*: with an account of their rise, progress and construction. Written with the aid of the records of the India Office. London: E. and F.N. Spon, 1868.

Davis, Lance E. and Robert A. Huttenback. *Mammon and the Pursuit of Empire: The Economics of British Imperialism.* Abridged edn; Cambridge: Cambridge University Press, 1988.

De Cosson, A.F.C. 'The Early Days of the East Indian Railway. A Side-Light on the Mutiny', *Bengal Past and Present,* v (January–June 1910), pp. 261–72.

Derbyshire, I.D. 'Economic Change and the Railways in North India, 1860–1914', *Modern Asian Studies,* 21:3 (July 1987), pp. 521–45.

Dewey, Clive (ed.). *The State and the Market: Studies in the Economic and Social History of the Third World.* Riverdale: The Riverdale Company, 1987.

Dewey, Clive J. (ed.). *Arrested Development in India: The Historical Dimension.* Riverdale, Maryland: The Riverdale Company, 1988

Dickinson, John. *Remarks on the Indian Railway Reports Published by the Government and Reasons for a Change of Policy in India.* London: P.S. King, 1862.

Dossal, Mariam. 'Henry Conybeare and the politics of centralised water supply in mid-nineteenth century Bombay', *IESHR,* 25:1 (January–March 1988), pp. 79–96.

Dowden, T.F. *Notes on Railways.* Bombay: Education Society's Press, Byculla, 1873.

Dulai, Surjit S. and Arthur Helweg (eds). *Punjab in Perspective: Proceedings of the Research Committee on Punjab Conference, 1987.* East Lansing: Asian Studies Center, Michigan State University, 1991.

Enthoven, R.E. *The Tribes and Castes of Bombay.* 3 vols. Bombay: 1921–3.

Friedman, Andrew L. *Industry and Labour: Class Struggle at Work and Monopoly Capitalism.* London: Macmillan, 1977.

Fuchs, Stephen. *At the Bottom of Indian Society: The Harijan and Other Low Castes.* New Delhi: Munshiram Manoharlal, 1981.

(Furnivall, W.C.) *Rajputana State Railway: Report on Construction by the Engineer-in-Chief.* Simla: Government Central Branch Press, 1875.

Gadgil, D.R. *Origins of the Modern Indian Business Class: An Interim Report.* New York: Institute of Pacific Relations, 1959.

Goodman, David and Michael Redclift. *From Peasant to Proletarian: Capitalist Development and Agrarian Transitions.* Oxford: Basil Blackwell, 1981.

Gorze, Andre (ed.). *The Division of Labour: The Labour Process and Class-*

struggle in Modern Capitalism. Atlantic Highlands, N.J.: Humanities Press, 1976.

Guha, Ramachandra. *The Unquiet Woods.* Ecological Change and Peasant Resistance in the Himalaya. Berkeley: University of California Press, 1990.

Guha, Ranajit (ed.). *Subaltern Studies*, vols. 1–6. Delhi: Oxford University Press, 1982–9.

Gustafson, W. Eric. 'The Gift of an Elephant? The Indian Guaranteed Railways, 1845–1870', unpublished paper presented at the annual meeting of the Association for Asian Studies, March 1971.

Hari Rao, P. *The Indian Railways Act* (IX of 1890) with all amendments and case-law up to date, full commentaries, four appendices and an Introduction. 2nd edn, Madras: Madras Law Journal Office, 1949.

Harrison, John. 'The Records of Indian Railways: A Neglected Resource', *South Asia Research*, 7:2 (November 1987), pp. 105–21.

Hawke, G.R. *Railways and Economic Growth in England and Wales, 1840–70.* Oxford: Clarendon Press, 1970.

Headrick, Daniel R. *The Tentacles of Progress.* Technology Transfer in the Age of Imperialism, 1850–1940. New York: Oxford University Press, 1988.

Hearn, Major G.R. and A.G. Watson. *Railway Engineers' Field Book.* A practical manual for Engineers in the East containing directions for the conduct of a Railway Survey, Instructions on the Tacheometer with reduction tables, Tables of Latitude and Departure, full curve tables for every minute of arc for all angles fron 0 o to 120 o and many others; also details for Transition curves. Calcutta: By the authors, 1913.

Hehir, Patrick. *Malaria in India.* London: Oxford University Press, 1927.

Helps, Arthur. *Life and Labours of Mr. Brassey, 1805–72.* London: Bell and Daldy, 1872.

Helps, Arthur. *Life and Labours of Mr. Brassey, 1805–72*; reprint edn, New York: Augustus M. Kelley, 1969.

Hirschman, Albert O. *Exit, Voice, and Loyalty: Responses to Decline in Firms, Organizations, and States.* Cambridge: Harvard University Press, 1970.

Holmstrom, Mark (ed.). *Work for Wages in South Asia.* New Delhi: Manohar, 1990.

Huddleston, G. *History of the East Indian Railway.* Calcutta: Thacker, Spink & Co., 1906.

(Huddleston, G.) 'The Opening of the East Indian Railway', *Bengal Past and Present*, II: 1 (January–July 1908), pp. 55–61.

Huddleston, George. *History of the East Indian Railway* (part II). Bristol: St. Stephen's Bristol Press, Ltd., 1939.

Hughes, H.C. 'India Office Railway Records', *Journal of Transport History*, VI:4 (November 1964), pp. 241–8.

Hughes, H.C. 'The Scinde Railway', *Journal of Transport History*, 5:4 (November 1962), pp. 219–25.

Hughes, J.R.T. *Fluctuations in Trade, Industry and Finance: A Study of British Economic Development.* Oxford: Clarendon Press, 1960.

Hurd, John. 'Railways and the Expansion of Markets in India, 1861–1921', *Explorations in Economic History*, 12:3 (July 1975), pp. 263–88.

Ichioka, Yuji. 'Japanese Immigrant Labour Contractors and the Northern Pacific and the Great Northern Railroad Companies, 1898–1907', *Labour History*, 21:3 (Summer, 1980), pp. 325–63.

Jenks, Leland H. *The Migration of British Capital to 1875.* 1927; reprint edn, London: Nelson, 1971.

Joyce, Patrick. *Work, Society and Politics: The Culture of the Factory in Later Victorian England.* New Brunswick, N.J.: Rutgers University Press, 1980.

Joyce, Patrick. 'Labour, capital and compromise: a response to Richard Price', *Social History*, 9:1 (January 1984), pp. 67–76.

Joyce, Patrick (ed.). *The Historical Meanings of Work.* Cambridge: Cambridge University Press, 1987.

Karaka, Dosabhai Framji. *History of the Parsis including their Manners, Customs, Religion, and Present Position.* 2 vols. London: Macmillan and Co., 1884.

Kaw, Mushtaq Ahmad. 'Some aspects of begar in Kashmir in the sixteenth to eighteenth centuries', *IESHR*, 27:4 (October–December 1990), pp. 465–74.

Kay, Geoffrey. *The Economic Theory of the Working Class.* New York: St. Martin's Press, 1979.

Kellett, John R. *Railways and Victorian Cities.* London: Routledge & Kegan Paul, 1979.

Kerr, Ian J. 'Constructing Railways in India—an estimate of the numbers employed, 1850–80', *IESHR*, 20:3 (July–September 1983), pp. 317–39.

Kerr, Ian J. 'Working Class Protest in 19th Century India. Example of Railway Workers', EPW, XX:4 (26 January 1985), PE34–PE40.

Kooiman, Dick. *Bombay Textile Labour: Managers, Trade Unionists and Officials 1919–1939.* New Delhi: Manohar, 1989.

Krishnamurty, J.K. (ed.). *Women in Colonial India: Essays on Survival, Work and the State.* Delhi: Oxford University Press, 1989.

Krishnamurty, Sunanda. 'Real Wages of Agricultural Labourers in the Bombay Deccan, 1874–1922', *IESHR,* XXIV:1 (January–March 1987), pp. 81–98.

Kumar, Dharma. *Land and Caste in South India: Agricultural Labour in the Madras Presidency during the Nineteenth Century.* Cambridge: Cambridge University Press, 1965.

Kumar, Dharma (ed.). *The Cambridge Economic History of India.* vol. 2: *c.* 1757–*c.* 1970. Cambridge: Cambridge University Press, 1983.

Kumar, Nita. *The Artisans of Banaras: Popular Culture and Identity, 1880–1986.* Princeton: Princeton University Press, 1988.

Landes, David S. *The Unbound Prometheus: Technological Change and Industrial Development in Western Europe from 1750 to the Present.* Cambridge: Cambridge University Press, 1969.

Latif, Syad Muhammad. *Lahore,* Lahore: New Imperial Press, 1892.

Lebra, Joyce, Joy Paulson and Jana Everett (eds). *Women and Work in India: Continuity and Change.* New Delhi: Promilla & Co., 1984.

Lee-Warner, Sir William. *The Life of the Marquis of Dalhousie, K.T.* 2 vols. London: Macmillan and Co., 1904.

Lehmann, Fritz. 'Railway Workshops, Technology Transfer, and Skilled Labour Recruitment in Colonial India', *Journal of Historical Research,* xx:1 (August 1977), pp. 49–61.

Longridge, James A. *Report on the Calcutta and South Eastern Railway, from Calcutta to the River Mutla, and the Extension Eastwards to Dacca and the Burmese Provinces.* London: 1857.

Ludden, David. *Peasant History in South India.* Princeton: Princeton University Press, 1985.

Macgeorge, G.W. *Ways and Works in India.* Being an account of the Public Works in that country from the earliest times up to the present day. Westminster: Archibald Constable and Company, 1894.

Macpherson, W.J. 'Investment in Indian Railways, 1845–1875', *Economic History Review,* 2nd series. VIII:2 (1955), pp. 177–86.

Malik, M.B.K. *Hundred Years of Pakistan Railways.* Karachi: Ministry of Railways and Communications, Government of Pakistan, 1962.

Mandel, Ernest. *Late Capitalism.* London: NLB, 1975.

Marshall, John. *A Biographical Dictionary of Railway Engineers.* Newton Abbot: David and Charles, 1978.

Marx, K. and F. Engels. *The First Indian War of Independence 1857–1859.* Moscow: Foreign Languages Publishing House, n.d.

Marx, Karl. *Capital: A Critique of Political Economy.* vol. 1. Introduced by Ernest Mandel. Translated by Ben Fowkes. New York: Random House, Vintage Books, 1977.

Marx, Karl. *Capital: A Critique of Political Economy.* vols. 2 and 3. Introduced by Ernest Mandel. Translated by David Fernbach. New York: Random House, Vintage Books, 1981.

Marx, Karl. *Pre-capitalist Economic Formations.* Translated by Jack Cohen. Edited and with an introduction by E.J. Hobsbawm. New York: International Publishers, 1965.

Marx, Karl. *The Poverty of Philosophy.* New York: International Publishers, 1963.

McAlpin, Michelle Burge. *Subject to Famine: Food Crises and Economic Change in Western India,* 1860–1920. Princeton: Princeton University Press, 1983.

McLeod, W.H. 'Ahluwalias and Ramgarhias: Two Sikh Castes', *South Asia,* no. 4 (October 1974), pp. 78–90.

McLeod, W.H. *The Evolution of the Sikh Community: Five Essays.* Oxford: Clarendon Press, 1976.

Medick, Hans. 'Industrialization before industrialization? rural industries in Europe and the genesis of capitalism', *IESHR,* 25:3 (July–September 1988), pp. 371–84.

Middlemas, Robert Keith. *The Master Builders: Thomas Brassey; Sir John Aird; Lord Cowdray; Sir John Norton-Griffiths.* London: Hutchinson of London, 1963.

Mitchell, B.R. 'The Coming of the Railway and United Kingdom Economic Growth', *Journal of Economic History,* 24:3 (September 1964), pp. 315–36.

Moore, Wilbert E. and Arnold S. Feldman (eds). *Labour Commitment and Social Change in Developing Areas.* New York: Social Science Research Council, 1960.

Morris, David Morris and Clyde B. Dudley. 'Selected Railway Statistics For the Indian Subcontinent (India, Pakistan and BanglaDesh), 1853—1946–47', *Artha Vijnana,* XVII: 3 (September 1975), pp. 187–298.

Morris, Morris David. *The Emergence of an Industrial Labour Force in*

India: A Study of the Bombay Cotton Mills, 1854–1947. Berkeley: University of California Press, 1965.

Morris, Morris David. 'Caste and the Evolution of the Industrial Workforce in India', *Proceedings of the American Philosophical Society*, 104:2 (April 1950), pp. 124–33.

Morris, Morris David. 'Values as an obstacle to economic growth in South Asia: An historical survey', *Journal of Economic History*, 27:4 (December 1967), pp. 588–607.

Morris, Morris David. 'Modern Business Organization and Labour Administration. Specific Adaptations to Indian Conditions of Risk and Uncertainty, 1850–1947', EPW (6 October 1979), 1680–7.

Mukerjee, Radhakamal. *The Indian Working Class*. Bombay: Hind Kitabs, 1945.

Mukherjee, Hena. 'The Early History of the East Indian Railway, 1845–1879'. Ph. D. dissertation, University of London, School of Oriental and African Studies, July 1966.

Muslow, B. and H. Finch (eds). *Proletarianisation in the Third World*. London: Croom Helm, 1984.

Nanjundayya, H.V. and L.K. Ananthakrishna Iyer. *The Mysore Tribes and Castes*. 4 vols. Mysore: Mysore University, 1931–5.

Nigam, Sanjay. 'Disciplining and policing the 'criminals by birth', Parts 1&2, *IESHR*, 27: 2&3 (April–September 1990)

O'Brien, Patrick. *The New Economic History of the Railways*. London: Croom Helm, 1977.

O'Hanlon, Rosalind. 'Recovering the Subject. Subaltern Studies and Histories of Resistance in Colonial South Asia', MAS, 22:1 (February 1988), pp. 189–224.

Omvedt, Gail. 'Migration in Colonial India: The Articulation of Feudalism and Capitalism by the Colonial State', JPS, 7:2 (January 1980), pp. 185–212.

Panigrahi, D.N. (ed.). *Economy, Society and Politics in Modern India*. New Delhi: Vikas, 1985.

Parry. J.W. 'Some Features of Indian Railways', *Engineering Magazine* (July 1898), pp. 558–74.

Patel, Surendra J. *Agricultural Labourers in Modern India and Pakistan*. Bombay: Current Book House, 1952.

Pentland, H.C. 'The Development of a Capitalistic Labour Market in Canada', *Canadian Journal of Economics and Political Science*, 25:4 (November 1959), pp. 450–61.

Perera, G.F. *The Ceylon Railway: The Story of its Inception and Progress.* Colombo: The Ceylon Observer, 1925.

Peto, H. *Sir Morton Peto: A Memorial Sketch.* London: printed for private circulation by Elliot Stock, 1893, p. 119.

Pinney, Thomas (ed.) *Kipling's India: Uncollected Sketches 1884–88.* Basingstoke: 1986.

Pollard, Sidney. *The Genesis of Modern Management: A Study of the Industrial Revolution in Great Britain.* London: Edwin Arnold, 1965.

Prakash, Gyan. *Bonded Histories: Genealogies of Labor Servitude In Colonial India.* Cambridge: Cambridge University Press, 1990.

Price, Richard. 'The labour process and labour history', *Social History,* 8:1 (January 1983), pp. 57–75.

Price, Richard. *Labour in British Society: An Interpretative History.* London: Croom Helm, 1986.

Price, Richard. 'Conflict and co-operation: a reply to Patrick Joyce', *Social History,* 9:2 (May 1984), pp. 217–31.

Ranade, Rekha. *Sir Bartle Frere and His Times* (A Study of His Bombay Years, 1862–7). New Delhi: Mittal Publications, 1990.

Rao, Y. Saraswathy. *The Railway Board.* A study in administration. New Delhi: S. Chand & Co., 1978.

Ray, Bharati. 'The genesis of railway development in Hyderabad state: a case study in nineteenth century British Imperialism', *IESHR,* 21:1 (January–March 1984), pp. 45–69.

Reed, M.C. (ed.). *Railways in the Victorian Economy: Studies in Finance and Economic Growth.* Newton Abbot: David & Charles, 1969.

Richards, F. *Salem* (Madras District Gazetteers). vol. 1, part 1. Madras: Government Press, 1918.

Risley, H.H. *The Tribes and Castes of Bengal: Ethnographic Glossary.* 2 vols. Calcutta: Bengal Secretariat Press, 1891–2.

Robertson, Murray. 'The Railways of India. Their Policy and Finance', *The Nineteenth Century and After,* LXX (July 1911), pp. 84–103.

Rose, H.A., compiler. *A Glossary of the Tribes and Castes of the Punjab and North-West Frontier Province.* 3 vols. Lahore: Civil and Military Gazette Press, 1911–19.

Rothermund, Dietmar. *An Economic History of India from Pre-colonial Times to 1986.* New Delhi: Manohar, 1989.

Rungta, Radhe Shyam. *The Rise of Business Corporations in India, 1851–1900.* Cambridge: Cambridge University Press, 1970.

Russell, R.V. *The Tribes and Castes of the Central Provinces of India.* 1916; reprint edn. Oosterhout: Anthropological Publications, 1969.

Sabel, Charles and Jonathan Zeitlin. 'Historical Alternatives to Mass Production: Politics, Markets and Technology in Nineteenth-Century Industrialization', *Past and Present,* 108 (August 1985), pp. 133–76.

Saberwal, Satish. *Mobile Men: Limits to Social Change in Urban Punjab.* New Delhi: Vikas, 1976.

Sahlins, Marshall. *Culture and Practical Reason.* Chicago: University of Chicago Press, 1976.

Sahni, J. N. *Indian Railways One Hundred Years, 1853 to 1953.* New Delhi: Government of India, Ministry of Railways (Railway Board), 1953.

Samuel, Raphael. 'Workshop of the World: Steam Power and Hand Technology in mid-Victorian Britain', *History Workshop,* 3 (Spring, 1977), pp. 6–72.

Samuel, Raphael (ed.). *Miners, Quarrymen and Saltworkers.* London: Routledge & Kegan Paul, 1977.

Sanyal, N. *The Development of Indian Railways.* Calcutta: 1930.

Satow, Michael and Ray Desmond. *Railways of the Raj.* New York & London: New York University Press, 1980.

Sharma, S.N. *History of the Great Indian Peninsula Railway.* 2 vols. Bombay: 1990.

Shlomowitz, Ralph and Lance Brennan. 'Mortality and Migrant Labour En Route to Assam, 1863–1924', *IESHR,* 27:3 (July–September 1990), pp. 313–30.

Silver, Arthur W. *Manchester Men and Indian Cotton 1847–1872.* Manchester: Manchester University Press, 1966.

Silverberg, James (ed.). *Social Mobility in the Caste System in India: An Interdisciplinary Symposium.* The Hague: Mouton, 1968.

Simmons, C.P. 'Recruiting and Organizing an Industrial Labour Force in Colonial India: The Case of the Coal Mining Industry, *c.* 1880–1939', *IESHR,* 13:4 (1976), pp. 455–86.

Simmons, Jack. *The Railways of England and Wales 1830–1914.* vol. 1: *The System and Its Working.* Leicester: Leicester University Press, 1978.

Singh, Harbans and N. Gerald Barrier (eds). *Essays in Honour of Dr Ganda Singh.* Patiala: Punjabi University, 1976.

Stone, E. Herbert. *The Nizam's State Railway.* London: Murray and Heath, 1876.

Subrahmanian, K.K., D.R. Veena and B.K. Parikh. *Construction Labour Market: A Study in Ahmedabad.* New Delhi: Concept Publishing Company, 1982.

Sullivan, Dick. *Navvyman.* London: Coracle Books, 1983.

Thane, Pat, Geoffrey Crossick and Roderick Floud (eds). *The Power of the Past: Essays for Eric Hobsbawm.* Cambridge: Cambridge University Press, 1984.

Thompson, E.P. *The Making of the English Working Class.* Harmondsworth: Penguin Books, 1968.

Thompson, F.M.L. (ed.). *The Cambridge Social History of Britain 1750–1950.* vol. 2: *People and Their Environment.* Cambridge: Cambridge University Press, 1990.

Thompson, Paul. *The Nature of Work: An Introduction to Debates on the Labour Process.* Basingstoke: Macmillan, 1983.

Thorner, Alice. 'Semi-Feudalism or Capitalism? Nature and Significance of the Debate', EPW (4, 11, and 18 December 1982)

Thorner, Daniel. *Investment in Empire*: British Railway and Steam Shipping Enterprise in India 1825–1849. Philadelphia: University of Pennsylvania Press, 1950.

Thorner, Daniel. 'Great Britain and the Development of India's Railways', *Journal of Economic History*, XI:4 (Fall, 1951), pp. 389–402.

Thorner, Daniel. *The Shaping of Modern India.* Bombay: Allied Publishers, 1980.

Thurston, Edgar. *Castes and Tribes of Southern India.* 6 vols. Madras: Government Press, 1909.

Tucker, Richard. 'Forest Management and Imperial Politics: Thana District, Bombay, 1823–1887', *IESHR*, 16:3 (July–September 1979), pp. 273–300.

Tucker, Robert C. (ed.). *The Marx-Engels Reader.* 2nd edn, New York: W.W. Norton & Co., 1978.

Vicajee, Framjee. *Political and Social Effects of Railways in India.* London: 1875.

Villeroi, B. (ed.). *A History of the North Western Railway.* Lahore: For the author at the 'News Press', 1896.

Walker, Charles. *Thomas Brassey: Railway Builder.* London: Frederick Muller Ltd., 1969.

Wallerstein, Immanuel (ed.). *Social Change: The Colonial Situation.* New York: John Wiley & Sons, 1966.

Walrond, Theodore (ed.). *Letters and Journals of James, Eighth Earl of Elgin.* London: John Murray, 1872.

Warren, Bill. *Imperialism: Pioneer of Capitalism.* London: Verso, 1980.

Washbrook, David. 'South Asia, the World System, and World Capitalism', JAS, 49:3 (August 1990), pp. 479–508.

Washbrook, David. 'Progress and Problems: South Asian Social and Economic History, *c.* 1720–1860', MAS, 22:1 (1988), pp. 57–96.

Way, Peter. 'Shovel and Shamrock: Irish Workers and Labour Violence in the Digging of the Chesapeake and Ohio Canal', *Labour History,* 30:4 (Fall, 1989), pp. 489–517.

Westwood, J.N. *Railways of India.* Newton Abbot: David & Charles, 1974.

Williamson, Thomas. *Two Letters on the Advantages of Railway Communication in Western India,* addressed to the Right Hon. Lord Wharncliffe, Chairman of the Great Indian Peninsula Railway Company. London: Richard and John E. Taylor, 1846.

Wolpe, Harold (ed.). *The Articulation of Modes of Production: Essays from Economy and Society.* London: Routledge & Kegan Paul, 1980.

Wylie, William T. 'Poverty, Distress and Disease: Labour and the Construction of the Rideau Canal, 1826–32', *Labour/Le Travailleur,* 11 (Spring, 1983), pp. 7–29.

Yang, Anand A. 'Peasants on the move: a study of internal migration in India', *Journal of Interdisciplinary History,* 10:1 (Summer, 1979), pp. 37–58.

Youngson, A.J. (ed.). *Economic Development in the Long Run.* London: George Allen & Unwin Ltd., 1972.

Zalduendo, Eduardo A. *Libras Y Rieles.* Las inversiones britanicas para el desarrollo de los ferrocarriles en Argentina, Brasil, Canada e India durante el siglo XIX. Buenos Aires: Editorial El Coloquio, 1975.

Index

accidents 138, 141, 157–8, 162–3, 167
Addis, A.W.C. 123, 126, 145, 166
advances 51–4, 58, 98, 102, 118, 120–2, 124, 151, 168, 188
Arratoon, Ter (contractor) 64
Assam 2
Assam-Bengal Railway 102, 183

Bayley, Victor (engineer) 112–13
Beldar 110–12
Bengal-Nagpur Railway 12, 53, 75, 189, 198
Berkley, James John (engineer) 45, 59–60, 87, 98–9, 114–15, 120, 151, 156, 174, 180, 195
Bhopal State Railway 77
Bhore Ghat incline 4, 51, 55–6, 58, 60, 74, 98, 103–4, 109, 114, 119–20, 122, 126–7, 131, 161–6, 176–8, 184, 186, 198, 206–9
Bombay, Baroda and Central India Railway (BB&CIR) 31, 62, 70, 72, 77, 144–5, 151, 157, 176, 181, 192
Brandon & Co. (contractor) 47
Brassey, Thomas (contractor) 45, 48–9, 67, 69, 78–9, 151
Brassey, Paxton, Wythes & Henfrey (contractor) 48–53
Brassey, Wythes & Henfrey (contractor) 69
Brassey, Wythes & Perry (contractor) 78
Braverman, Harry xiii, 6–7, 169, 176, 193
Bray, James and Edwin (contractor) 62–3, 121, 178, 204
Bray, Joseph (contractor) 57
brickmaking 1, 3, 9, 101, 114, 129, 140, 143–8, 151, 155, 176, 189
bridges 128, 132–4
  Chakdara 158
  Chitravati 4
  construction 9, 31, 46, 113–14, 117, 121, 134–43, 152–6, 189, 198, 201, 207–8
  Curzon 207
  defective or failed 12, 33, 41, 58, 61–2, 69, 134–5, 144
  Dufferin 133, 136, 154, 157, 199
  EBR 137
  Empress 137, 140–1, 146
  Godaveri 154, 180
  Hooghly 167
  iron and steel 9, 135, 139–42, 156, 193
  Jubbulpore extension, EIR 141
  Kaisar-i-Hind 140, 151
  Kistna at Bezwada 108, 116–18, 152–6, 174
  Kurumnasa 103
  Lansdowne 25, 131, 133, 135, 141, 154, 157–8
  Nerbudda 135, 157
  Papagni 138
  PNSR 117, 140, 158, 181
  Sher Shah 101, 117, 140, 152–6
  Soane 36, 103
  SP&DR 134, 142
  Vellore 104, 177
British workmen 2, 22, 34, 117, 127, 130, 134, 138–40, 149–51, 162–3, 166–7, 174, 177, 197, 198–200, 209

Bruce, George Barclay (engineer) 70–1, 107, 119, 149
Brundell, Richard Shaw (engineer) 80
Brundell and Easton (contractor) 81
Brunton, John (engineer) 63–4, 76, 98–9, 116, 118, 173–4
Brunton, William (engineer) 63
Burawoy, Michael xiii, 9, 178, 192
Burn & Co. (contractor) 46–8

Calcutta and South-eastern Railway 31, 47, 94
canals 104–5
Canning Lord 18, 35–6, 103
capital-intensive construction 60, 87, 126, 132, 145
capitalism; *also* mode of production xi–xiii, 4–9, 16, 25, 52–3, 55, 68, 75–6, 83–4, 96, 104, 116, 120, 126–7, 152, 169–72, 185, 188, 190, 192–4
Carr, Mark W. (engineer) 80
caste and tribe 32, 55, 85–6, 90, 92, 96, 98, 100–19, 123
Chakrabarty, Dipesh 178
Chaman Extension of the IVSR 101, 162
Chandra, Bipan 17
Chevalier De Cortanze (contractor) 64, 66
child labour xii, 1, 85–9, 94, 105, 111–12, 114, 132, 156, 161, 175–6, 190
circulating labour 89–92, 97, 101–3, 106, 109–12, 118, 120–1, 126, 170–1, 173, 181, 190–1
Clowser (engineer) 60–1
Coates (contractor) 64, 66
commissariat arrangements 54, 151, 161
construction
  ballasting and plate-laying 130, 147–50, 188, 199
  building of stations, workshops etc. 130, 145–6, 188
  formation of the line 130–2, 146, 188
construction employment, estimates of *see* employment, estimates of
construction materials, provision and transport of 1, 3, 22–3, 33, 47, 64, 68, 81, 83, 85, 128–31, 135, 143, 147–8, 153–5, 190
construction problems and obstacles 15, 31–3, 35–6, 41, 47, 49–50, 134, 152, 182, 186–7, 193, 195, 202–3, 205–6
construction process 1–4, 9, 12–14, 16, 19, 20–2, 25, 31, 41, 68, 72, 81, 113, 127–8, 154, 184, 187–9, 192, 194, 196
contractors and contracts 2, 6, 7, 10–11, 18–19, 23–4, 27, 44–84, 92, 100, 105–6, 109, 112, 118, 122–3, 128–30, 145, 148, 151–2, 160, 163, 171, 173, 178–9, 185, 187–8, 190, 197, 200, 204
  disputes with 63, 66, 77–8, 121–2, 178–9
  failures 46–7, 56–7, 60, 62–3, 65–7, 74, 121, 123, 178, 204
  large contract system 23–4, 44–51, 56–9, 69–70, 73–5, 77, 79–80, 82–3, 130, 187
  petty 2, 5, 8, 13, 47–8, 50–3, 64–5, 70–1, 73–6, 84, 91, 102, 111, 113, 118–19, 121, 123–4, 126, 129–30, 140, 144, 147, 151, 173, 179, 187–9, 207
Cotton, Sir Arthur 17

Dalhousie, Lord 26, 29–30, 43
Danvers, Juland 19, 198
Das, Jumna (contractor) 64–5
Davidson, Edward 86, 132, 175
Davis and Huttenback 18
Deonath (contractor) 65
departmental system of construction 24, 53–4, 62, 70–7, 82–3, 123, 129, 145, 152, 179–80, 187
Dickens, Major-General 76, 78

disease 12, 32–3, 97–8, 102, 124, 155, 159–66, 191, 204, 206, 210
Dorabji, Jamsetji (contractor) 59–62, 65, 67–8, 187
Duckett and Stead (contractor) 56
Dutt, R.C. 17

earthworking, earthworkers 9, 10, 54, 86–90, 102–3, 105–13, 122–3, 129, 131–3, 160, 170, 174–5, 181, 189–90, 202
East Coast State Railway 12, 117, 152, 154
East Indian Railway (EIR) 17, 23, 27, 29, 31, 35–6, 42, 45–7, 50, 70, 79, 86, 93–4, 97, 99, 103–4, 113–14, 125, 141, 143, 167, 183, 197, 202–3, 208
Eastern Bengal Railway (EBR) 31, 48–51, 99–100, 125, 137, 173
Easton, John, M. (engineer) 80
Elgin, Lord 90–1, 110
Employers and Workmen (Disputes) Act (X) of 1860 51, 184
employment, enumerations and estimates of 40–2, 85, 98, 100, 113, 126–7, 140–1, 144–5, 148–9, 189–90, 197–210; (tables) 211–26
engineering colleges 34, 52
engineering journals, technical literature etc. 34, 126, 137, 139, 165
engineers 2, 6, 7, 10–12, 14, 21–5, 32–4, 37–8, 46–7, 70–3, 77–9, 81–4, 88, 98, 109–10, 117–18, 121, 123, 127–31, 133–5, 138–9, 143, 151–3, 155–6, 160, 173–4, 179–85, 187–90, 193, 195, 197–8, 201, 208
engineers' careers 74, 79–82, 153
engineers, levels and varieties of 2, 21, 23–5, 70, 79–80, 121, 127–30
Enthoven, R.E. 111

Famine Commission of 1880 43
famine labour xiii–xiv, 205
family labour 86–91, 112, 118, 124, 175–6, 188, 190
Faviell, William Frederick (contractor) 32, 56, 61
female labour xii, 1, 85–9, 94, 105, 111–12, 114, 119, 132, 144, 156, 161, 175–6, 190
Firbank, Joseph (contractor) 45
Forde, A.W. (engineer) 72, 97
Fowler, Henry (contractor) 32, 183
free and/or bonded labour 4–5, 53–4, 71, 73, 92, 96, 170, 173, 191
Frere, Bartle 4–5, 71, 92–3, 96

gangers (maistry, mate, muccadam, sardar) 2, 52–4, 67, 70, 72, 98–9, 107, 111, 119–23, 126, 168, 177, 179, 188
George, E. Monson 87, 132
Ghose, Preonath (engineer) 153
Glover, Thomas Craigie (contractor) 58, 77–8, 90, 95, 109–10, 121, 191, 199
Going, Thomas Hardinge (engineer) 86, 107–8, 174–5
Government 5, 10, 16–24, 33, 43, 50, 93, 150, 167, 177, 183–5, 187, 192
  control of construction process 18, 24, 32–3, 83, 93, 187
  Public Works 21, 24–5, 33–4, 50, 66–7, 75–6, 87, 93, 100, 106, 109, 120, 125, 199
Great Indian Peninsula Railway (GIPR) 17, 27–8, 31–2, 45, 48, 55–60, 62, 70, 77, 79, 81, 98, 109, 126, 144, 157, 198, 203, 207–8
Great Southern Railway of India (GSIR) 30, 69–70, 80
Guha, Ranajit 179

Hamilton, Brown & Co. (contractor) 81
harvest time 91–5, 110, 124

Henfrey, Charles (contractor) 48–50, 52, 54–5, 69, 78, 95, 99–101, 172, 183
Holkar State Railway 150
Hood, Winton and Mills (contractor) 57
Hookum Singh (contractor) 65
Hughes, J.R.T. 205–6
Hunt & Co. (contractor) 47
Hunt, Bray & Elmsley (contractor) 46

India Office 2, 22
Indian Midland Railway 76–7, 204
Indianization of supervisory and skilled workforce 117, 124, 129, 138–9, 149–50, 190, 198–9
Indus Valley State Railway (IVSR) 43, 76, 133, 141, 158, 181, 198, 209

Jackson (contractor) 45
Jubbulpore extension of the EIR 79–81, 83, 95, 174, 197

Kandahar State Railway 209
Kaw, Mushtaq Ahmad 94
Kennedy, J.P. 26, 70
Ker, C.B. (engineer) 58
Khyber Railway 100
Kipling, Rudyard 140–1, 151

labour process xi–xiii, 1, 3, 6–10, 13–14, 32, 52, 72, 89, 127, 130, 133, 143–5, 150, 152, 168–9, 189, 192, 194
  formal subsumption of labour xii, 7–9, 113, 150, 152, 190, 193–4
  real subsumption of labour xii, 7–9, 152, 193
labour, Indian xii, 2, 5–6, 85–8, 186
  child *see* child labour
  control of 10, 14, 72, 87, 101, 104, 119–20, 122–3, 126, 145, 147, 150, 152, 154, 156, 164, 177, 184–5, 189, 191, 194
  day hired 55, 71–2, 122–3, 129
  extra-economic relationships of 5, 10–11, 51, 53, 61, 91, 93–4, 113, 118–20, 123, 171, 183, 188, 194
  family *see* family labour
  impressed 93–4, 109, 183
  increased spatial mobility 33, 101–3, 117, 120–1, 125, 138, 160, 171, 191
  markets 4, 71, 73, 76, 91–2, 97, 99–103, 120–1, 124–6, 152, 171, 182–3, 191, 194
  recruiting and/or retaining 2, 12–13, 31, 33, 49–55, 60–3, 69, 71, 76, 85, 88, 90–1, 93–5, 97–104, 107–10, 113, 117–18, 120–6, 141–2, 151, 153, 160–1, 166, 171–2, 179, 181, 188, 190–1, 196, 200–5, 209, *see also* recruitment of labour, social and geographical sources
  resistance of *see* worker resistance
  skilled xii, 13, 33–4, 85, 88, 91, 97–8, 100–2, 108, 113–18, 121, 124, 130, 138, 139–41, 149, 153, 156, 173, 181, 184, 190, 194–5, 199
  training 34, 108–9, 115–17, 124, 130, 138–9, 142, 149–50, 191
  unskilled 2, 13, 85, 87–8, 90–1, 95, 97–8, 100–2, 103–4, 113–16, 118, 121, 124, 129, 131, 140, 156, 190
  women *see* female labour
  work organization 12, 107, 113–15, 119, 123, 127, 129–31, 133, 138, 141, 143–5, 147–9, 152, 156, 169, 188–9, 205
  working/living conditions 12, 56, 61, 88, 98, 133–4, 137–8, 140, 155–68, 176, 191–2
Lal, Chota (contractor) 65
Lawrence, John 63–4, 66, 68–9
LeMesurier, Charles (engineer) 79
LeMesurier, Henry Peveril (engineer) 79–80, 83, 97
Ludden, David 96

Macgeorge, G.W. 13; 132, 189
Madras Railway (MR) 4, 23, 30, 69–70, 86, 93, 107, 149, 201, 209
management and managerial authority xi–xiii, 2–4, 7, 9–10, 12, 20–6, 52, 61, 64, 72, 81–4, 113, 115, 119, 123, 127, 129, 150, 154–6, 168, 174, 177–8, 183, 187–9, 191–5
Marx, Karl xi, xiii, 6–9, 14, 96, 156, 192–3
Medley, J.G. 111
Mettur Dam 10
Monghyr Tunnel 34, 103, 133, 202
Moplahs 86
mortality 32–3, 97–8, 157–63, 167, 191, 195
Mughal India 147
Mutiny and civil disturbances, 1857 58 35–7, 125, 184

Naoroji, Dadabhai 17
Napier, Lord 4
Nargunjoo Pass 103
navvies 13, 45, 90, 105–7, 109–12, 132, 168
Nelson & Co. (contractor) 35, 47
Nicoll, George (engineer) 80
Nizam's State Railway 123, 126, 171
Norris & Co. (contractor) 46–7
Norris & Weller (contractor) 57
North Western Railway (NWR) 76, 83, 117, 140, 152
Nunias (Loonias, Luniyas) 100, 112–13, 190

Ods (Odh, Ud) 110–11, 190
Omerdeen (contractor) 64–5
Oudh & Rohilkhand Railway 76, 199

Pallas 96
Parsis 59–61
Paxton, Sir Joseph (contractor) 48
Perry (contractor) 78
Peto, Sir Samuel Morton (contractor) 45, 82
Peto & Betts (contractor) 69
Pouchepadass, Jacques 96
Prakash, Gyan 96
Prem Singh (contractor) 65
Price, Richard 3, 14
princely states 81
proletarians and proletarianization xiii, 92–3, 96, 170–3, 181, 192, 194–5
Punjab Northern State Railway (PNSR) 76, 79, 117, 140, 158, 181, 204, 208

railways,
financing 1, 2, 4, 6, 18–19, 36–7, 44, 187–8
gauge 21, 26
guarantee 17–18, 128–87
initial line openings 27–8, 30–1
materials for, ordered in Britain 2, 22, 131, 135, 141, 154
mileage, construction and open line 1, 38–41, 43, 126, 186–7, 195, 208
promotion 16–17
state *see* state railways
travel by 186
Rajputana State Railway 25, 77
Ram, Mela (contractor) 64
Ramgarhias 116
reconstruction employment 41, 191, 210
record keeping 128
recruitment of labour, social and geographical sources of 85–126, 142, 155, 163–4, 171, 177, 181, 190, 200–5
Rendel, Sir Alexander 25
Ripon, Lord 43
river training 134, 153, 156
riveting 117, 141–3, 152–4, 174, 181, 205
Rose, H.A. 111
Ryan (contractor) 46

Santhal rebellion 35, 47, 179
science and technology 8, 11–12, 108, 132, 156, 193

security of railways, British concern for 37
Sgardelli 167
shift work 133, 162
Simmons, Jack 44
Sind, Punjab and Delhi Railway (SP&DR) 31, 62–5, 68–9, 95, 98–9, 121, 203–4, 209
Sindia State Railway 77
sleepers 33, 131, 147–8
Smith, Adam 6
Sobhan, Sheik 167, 195
Southern Railway of India (SIR) 30, 75, 87, 95, 118, 121–2, 180
Spencer, Charles Innes (engineer) 79–80
Spring, Francis J.E. (engineer) 108, 116, 153–6
state railways 20–1, 23, 43, 74–5, 83, 94, 128, 187
stations 37, 64, 66, 94, 113, 128, 146, 148, 186
Stephenson, George 12, 156
Stephenson, Robert 12, 80
Stevens, F.W. 146
stone-working, also masons 98, 101, 106–11, 113–18, 122, 129, 140, 179, 201
Strachey, General 75
Sudan, plate-laying in 150
Sultan, Muhammad (contractor) 64–8, 123, 187
survey 27, 100, 128, 170–1
Swan & Musgrave (contractor) 69
Syme, D. 158

Tapti Valley Railway 95
technology: forms, adaptations, innovations & transfer 11–13, 31, 116, 130, 132–9, 142–4, 151–4, 173–5, 193, 205
Thal Ghat incline 55–6, 60, 158, 164, 177, 186, 198, 206–9
Theroux, Paul 1
Thurston, Edgar 106
tools and machines 3, 11–12, 44, 98, 108, 116, 130–3, 136–8, 140–5, 152–4, 156, 173–5, 183, 189, 193, 205
Tredwell, Alice (contractor) 56
Tredwell, Solomon (contractor) 56
truck system and the Tommy Shop 151, 168
tunnels 34, 53, 103, 132–4, 162, 198–9, 207
Turnbull, George (engineer) 97, 203

Vidyant, Babu Ram Gopal (engineer) 199
violence 104, 166–8, 176–8

Wadia, Dadabhoy Pestonji 59
wages: amounts and forms of payment 10, 36, 50, 54, 62, 87, 100–2, 112–13, 119, 123, 125, 129, 150–2, 168, 170, 173, 175–81, 184, 190, 192, 202, 204
Waring Brothers & Hunt (contractor) 69, 79–80, 82
well-sinking for bridge foundations 135–8, 154, 174, 180, 205
West, Arthur (engineer) 108–9, 131, 174
West of India Portuguese Railway 132, 157
work, conduct of 107–8, 131–3, 139–41, 144–5, 148–9, 155, 174, 188, 191, 193
worker resistance 13–14, 104, 122, 168–85, 189, 191–2, 195
  other than strikes 50, 104, 122, 170–80
  strikes 122, 170, 176, 180–4
Wudders (Oddar, Odde, Vadda, Vaddar, Waddar) 105–11, 113–14, 116, 119, 121, 148, 174, 190–1
Wythes, George (contractor) 48, 57, 82
Wythes & Jackson (contractor) 57–8, 60